AA003221

Management Division 2015

Core Programming Area at the 2015 AIChE Spring Meeting and 11th Global Congress on Process Safety

Austin, Texas, USA
26-30 April 2015

ISBN: 978-1-5108-0705-1

Printed from e-media with permission by:

Curran Associates, Inc.
57 Morehouse Lane
Red Hook, NY 12571

Some format issues inherent in the e-media version may also appear in this print version.

Copyright© (2015) by AIChE
All rights reserved.

Printed by Curran Associates, Inc. (2015)

For permission requests, please contact AIChE
at the address below.

AIChE
3 Park Avenue
New York, NY 10016-5991

Phone: (203) 702-7660
Fax: (203) 775-5177

www.aiche.org

Additional copies of this publication are available from:

Curran Associates, Inc.
57 Morehouse Lane
Red Hook, NY 12571 USA
Phone: 845-758-0400
Fax: 845-758-2634
Email: curran@proceedings.com
Web: www.proceedings.com

TABLE OF CONTENTS

409505 Chemical Engineering: Is This the Golden Age?

Abstract:

Schlanger will offer his insights on the current opportunities in the chemical process industries in his talk entitled "Chemical Engineering: Is This the Golden Age?" on Monday morning, April 27. At the conclusion of his lecture, Schlanger will receive the AGILE Award in recognition of his leadership at CEVA Group, a $9 billion logistics supplier to industry worldwide; LyondellBasell, a major player in the plastics, chemicals, and refining industries; Arco Chemical; and at the other companies where he has held leadership positions.

Copyright © American Institute of Chemical Engineers. All rights reserved.

401746 Industrial Perspective on Licensing and Credentialing

Abstract:

In the USA large and small industrial organizations provide many engineering services, and they design, build and commission facilities in North America and globally. Standards and practices for continuing professional competency (CPC) vary for all aspects of the profession including engineering skills, ethics and social responsibility. ABET has increasing taken a role of adding these types of requirements to undergraduate chemical engineering programs. The recent policy statement by AIChE and ASME on the Industrial Exemption could be considered a compromise among industry and society and the engineers. This presentation will give one view of a licensed PE who has worked in organizations as small as two engineers to the largest ones in the world doing research to ECP to plant operations. Often, the hurdles to an engineer's obtaining a license are the exams to show engineering skills competency.

Copyright © American Institute of Chemical Engineers. All rights reserved.

399537 Sustainability Key Performance Indicators

Abstract:

Sustainability and climatle action require data and metrics to successfully measure performance towards goals and targets. The City of Austin has been pursuing climate action goals since 2007 and city-wide sustainabiltiy metrics since 2010. This presentation will explore sustainability and climate metrics, data collection, analysis, management, reporting, and successfully using this information to drive performance and change. With 4-5 years of data from city operations, which include 250 buildigs, 12,000 employees, 5,000 vehicles, a municipally owned water and power utility, as well as community wide sustainability outcomes, the amount of data and opportunities to communicate data are immense.

Copyright © American Institute of Chemical Engineers. All rights reserved.

400440 Making Sustainability Tangible: Resilience, Behavior Change, and Risk

Abstract:

When corporations push forward on projects based on cost-benefit analyses that weigh social, economic, and environmental factors; the economic element of the triple bottom line often outweighs the other two components. Why is this? The answer is.....substance. When it comes to sustainability, managers respond most effectively when the outcomes move beyond the qualitative to the quantitative. In addition, sustainability has far reaching impacts on management decisions when approached through the lens of resilience and externalities. This presentation will explore scenarios and pose key questions about the critical importance of not only heavily incorporating all three components of the sustainability pyramid, but also how enhancing the social and environmental facets can cause unexpected paradigm shifts. Case studies will include disaster recovery and resilience, green office/operations practices, and employee engagement.

Copyright © American Institute of Chemical Engineers. All rights reserved.

398145 How Sustainability Is Infiltrating Business Planning

Abstract:

The executive responsible for sustainability within an organization often serves in the role of management consultant, coach and advocate for sustainability practices throughout the organization. While many may find their influence limited to responding to sustainability questionnaires, others are gaining traction infiltrating sustainability into the myriad aspects of business planning. In many cases the bright line between a sustainabilit-driven activity, and general business practice is dimming. Drawing on case studies this talk will examine examples where sustainability practices are mainstreaming...and why. Examples will include corporate activities around water stewardship, product innovation, and supply chain management.

Copyright © American Institute of Chemical Engineers. All rights reserved.

Improving Sustainability:

Employees Can Design/Build the Car, but Management Holds the Car Keys

Paul Westbrook
Sustainable Development Manager, LEED AP
Texas Instruments

April 2015

TEXAS INSTRUMENTS

Outline

- Sustainability Defined
- Technology vs Technique (Process)
- Integrated Design
- TI Case Study
- TI Results

- Abstract: The adoption of more sustainable practices hinges on two key areas. One is the technology development that allows for improved processes. The other, which is more difficult, requires changes in our behavior, habits, and even our culture. Management sets the expectations and examples for the company culture and can have a tremendous influence on the speed and depth of sustainable practice adoption. Texas Instruments (TI) has implemented a variety of sustainability initiatives over the years. This presentation will explore the key factors that led to successful initiatives and review the results of those initiatives. Some of TI's successes have been building the first LEED Gold semiconductor manufacturing facility in the world and cutting chip production energy intensity in half in just over 8 years.

TEXAS INSTRUMENTS

Sustainability Defined

- Official Definition

Brundtland Commission of the United Nations, 1987

Simply stated . . .
The balance of people, profit, and the planet

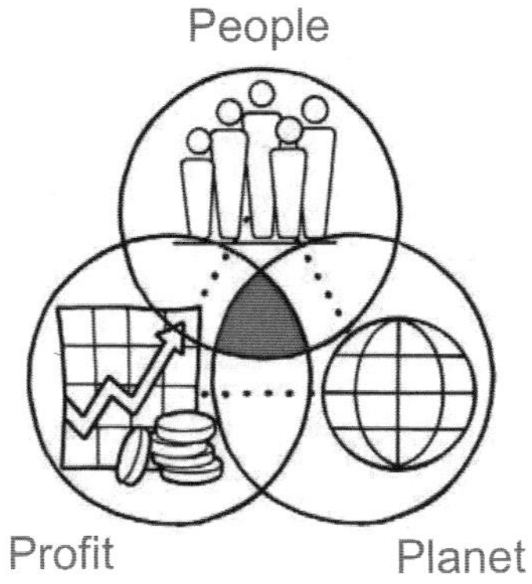

TEXAS INSTRUMENTS

Technology vs Technique

Technology

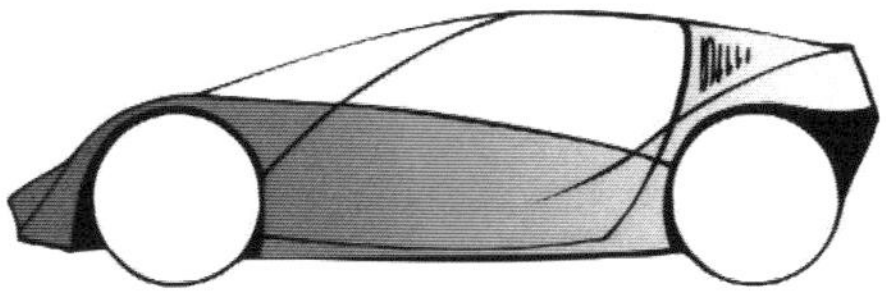

Machinery and equipment developed from the application of scientific knowledge

Technique (or Process)

A way of carrying out a particular task
How we organize our work flow/teams

TEXAS INSTRUMENTS

Technology vs Technique

Technology

- LED lights
- VFD motors
- Efficient / optimized systems
- Energy dashboards

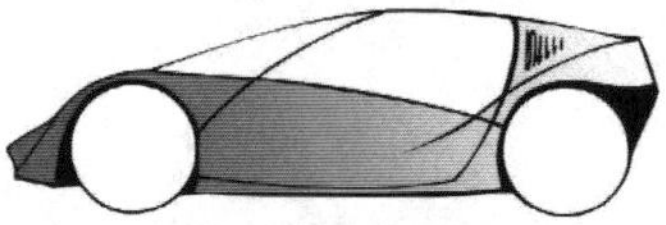

Technique

- Behavior
- Habits
- Culture
- Ethics
- Trust
- Support
- Organization

Texas Instruments

Management Style Spectrum

Servant Leadership

- Asking / Listening
- People oriented / Empathy
- Awareness
- Persuasion
- Planning / Foresight
- Stewardship
- Motivating
- Creative
- Constructive
- Focus on people's strengths

Top Down Micromanagement

- Telling
- Task-oriented
- Controlling
- Directing
- Deadline driven
- Done at all costs
- Driver
- Risk-averse
- Authoritarian
- Focus on "fixing" people's weaknesses

Texas Instruments

The Design Process

"If it ain't broke, don't fix it."

Note: It's broken

TEXAS INSTRUMENTS

Integrated Design – Building Example

An iterative, non-linear process

Typical Linear Design Process

1. Architect draws up design

2. Everyone else makes it work:
- High cooling load window locations
- No room for ductwork
- Excessive materials use / waste
- Poor daylighting

. . . Over budget,
then "value engineering" occurs

Integrated Design Process

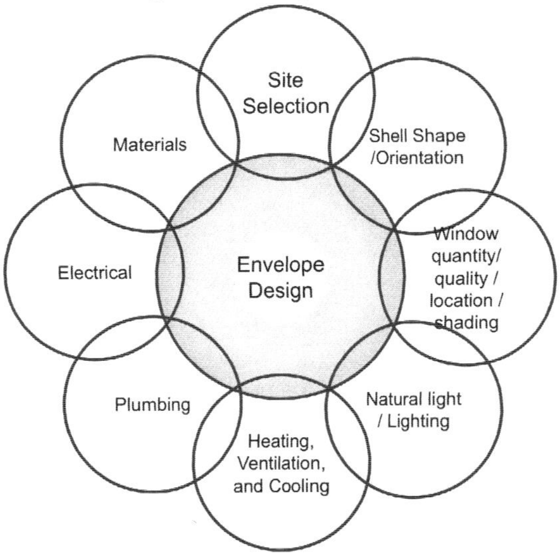

TEXAS INSTRUMENTS

Integrated Design

Value Engineering is Neither
- Amory Lovins, RMI

TEXAS INSTRUMENTS

Integrated Design: Loops

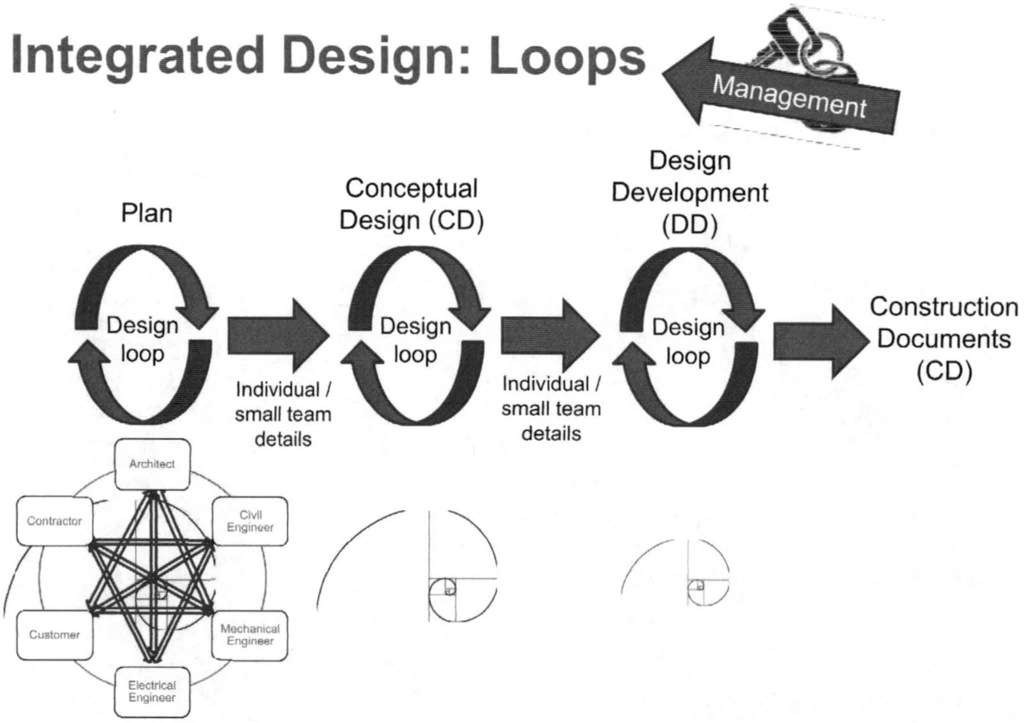

TEXAS INSTRUMENTS

Texas Instruments Example

TEXAS INSTRUMENTS

TI Business

2014 TI Revenue: $13.0 billion

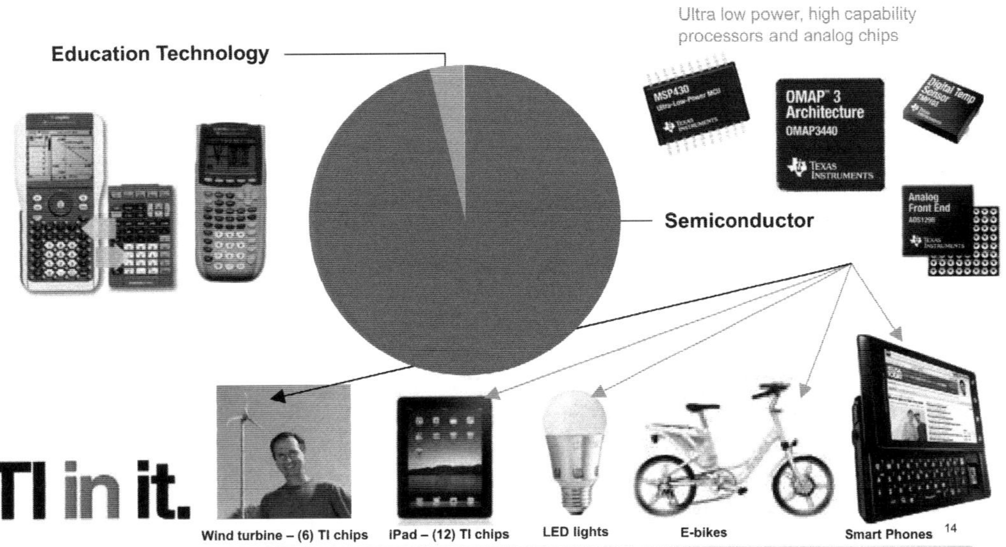

TI in it.

14

TEXAS INSTRUMENTS

TI's Global Presence

Approximately 32,000 employees worldwide

34 large sites
22 million gsf

Manufacturing, design, or sales operations in more than 30 countries serving over 80,000 customers

15

TEXAS INSTRUMENTS

What is a wafer fab?

A very big, clean facility . . .

Total space:
1,100,000 gsf (102,000 m2)
Clean room space:
220,000 sf (20,400 m2)

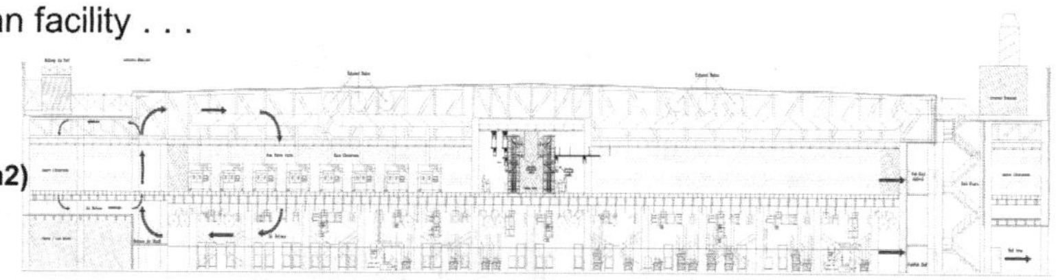

. . . that fabricates very small chips on large silicon wafers

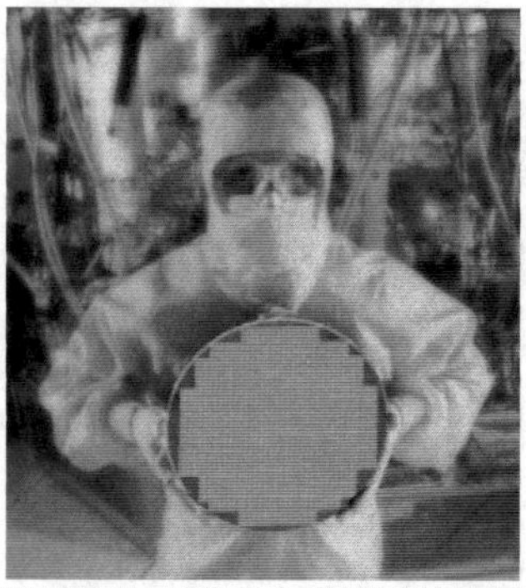

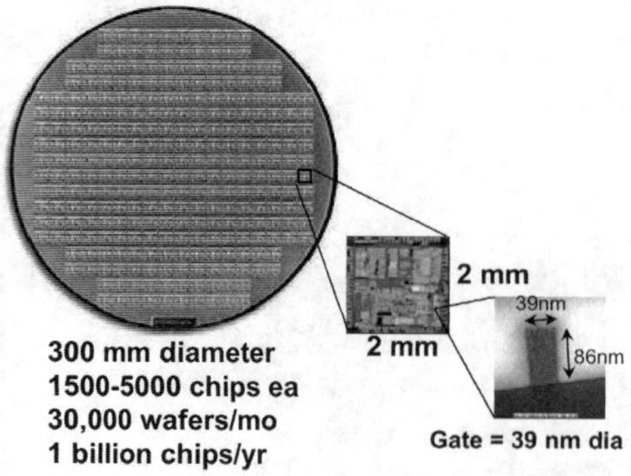

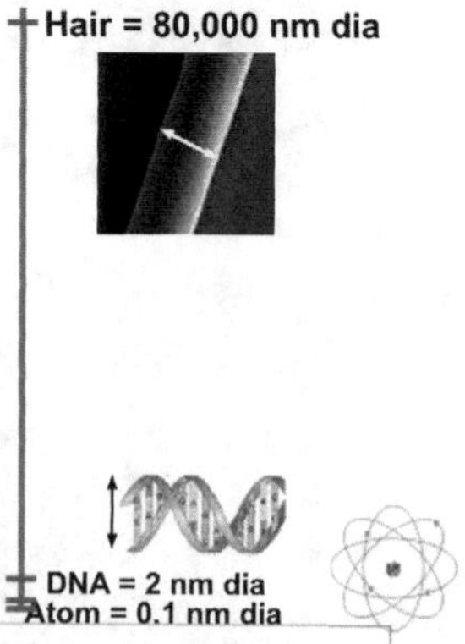

TEXAS INSTRUMENTS

Measurement

- Does anyone have a sustainable measuring device?
- Measurement is easier when broken into the major components:
  - Planet Factors
    - Energy
    - CO2
    - Emissions
    - Water
    - Materials
    - Waste

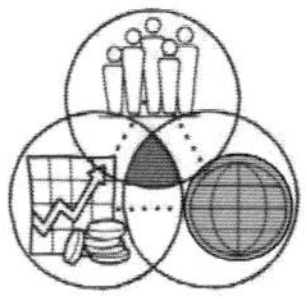

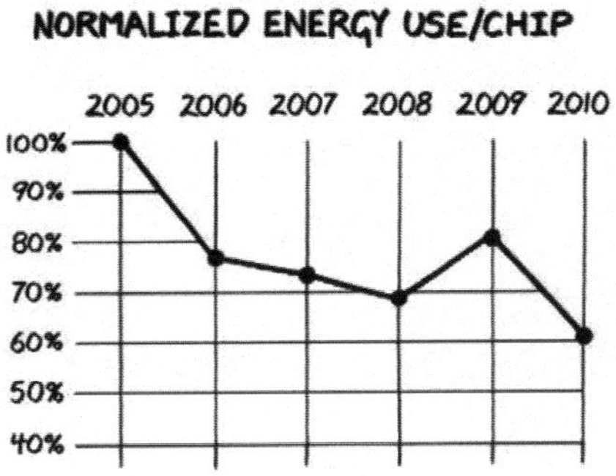

- Data can be normalized against square footage, units produced, weather, etc.
- Data is tracked over time and shared with employees to engage their assistance

17

Example Case: Semiconductor Fab

- 92-acre site
- 1.1 million square feet
- 220,000 square feet of cleanroom space
- Capacity for about 1,000 employees
- Main components:
  - Fab building (Fab=fabrication)
  - Support building
  - Central utility plant (CUP) building
  - Administration building

The Opportunity

- Very tight temperature and humidity requirements . . .
  - 70F+/-2 (21C+/-1) and 45% RH +/- 3%
- Combined with a large amount of exhaust and subsequent make up air . . .
  - 650,000 cfm (307 m3/sec) = 2 Macy's Kermit balloons per second

- Combined with the need to recirculate a large volume of air through the filters for cleanliness . . .
  - 4,400,000 cfm (2077 m3/sec) = 22 Goodyear blimps a minute

- Combined with hundreds of process tools with vacuum pumps, RF generators, and support equipment . . .
- Combined with extensive use of deionized (DI) water to rinse the wafers during processing . . .

Could lead to annual power consumption of 170,000 mWh (10,000 homes worth) and water consumption of 3 million gallons/day (6,000 homes worth). Annual utility bills could total $20M - $25M.

TEXAS INSTRUMENTS

Strategy

- Strategy Team - Fabscape
  - 4 strategy teams were formed in advance of project
  - Request made to add a 5th team - sustainability
  - Generated early white papers on a number of ideas

- Tour My House (enerjazz.com/house)
  - Invited 3 TI VP's to tour active/passive solar home
  - Low utility bills for "normal" house spurred interest

- Design Workshop
  - Teamed up with Rocky Mountain Institute (RMI)
  - Held 3-day design charrette to brainstorm ideas
  - Generated 15 "Big Honkin' Ideas" to carry forward along with a large list of other good ideas
  - Made a first pass at LEED score sheet

Rocky Mountain Institute
abundance by design™

TEXAS INSTRUMENTS

Project Management

- Sustainable Development Manager role assigned
  - Chief integrator of ideas and systems
  - The building was the system to be optimized
- Project Manager was fully supportive
  - Safety, Cost, Schedule, and Sustainability were front page
- Construction Contractor was the general contractor with the design firm working under them
  - Everyone on board from the start

Don't optimize the parts and pessimize the system

Cost Reduction – Friend or Foe?

- Challenge:

  - Reduce project costs per square foot by 30 percent from the previous semiconductor fab
- Response:
  - A LEED building does not have to cost more
  - Forced space efficiency
  - Forced us to question everything
  - Couldn't just copy previous design – had to innovate
  - All of this led to . . . . Engineering!

Integrated Design

You know you are on the right track when your solution for one problem accidentally solves several others.
- Amory Lovins, RMI

Texas Instruments

Integrated Design Example

Office Lighting

Standard cost = $125
op cost = $40/yr

Ergolight cost = $375
op cost = $25/yr

Simple payback = 16.7 years 

However, we need 30% fewer Ergolight fixtures. Simple payback down to 6.7 years.

Efficient lighting also saves cooling energy. Simple payback down to 6.0 years.

Typical Office Building Energy End Use - Texas

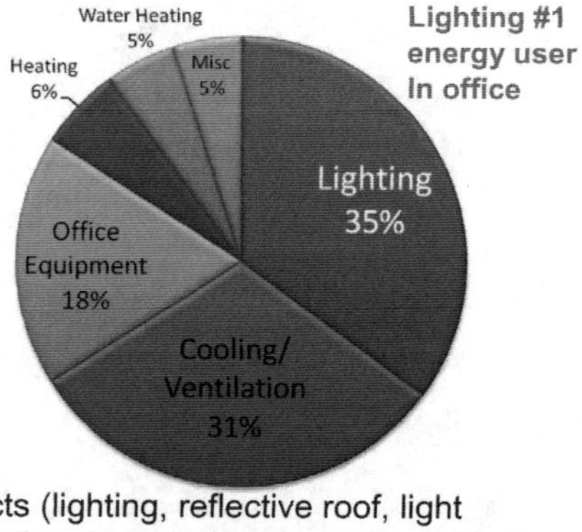

Add the contribution from dozens of similar projects (lighting, reflective roof, light shelves, sun shades, quality windows, extra insulation, better vacuum pumps, . .)

Enough cooling load disappears to avoid buying a $1M chiller . . . and the cooling tower, pumps, pipes, and even the space needed to install it.

Simple payback is now 0.0 years. The total net capital cost is the same, or even less, and the operating costs are lowered forever.

Texas Instruments

Integrated Design Example

But wait, it gets even better . . .

Ergolights are individually controllable by each employee
Natural daylighting has been shown to increase productivity

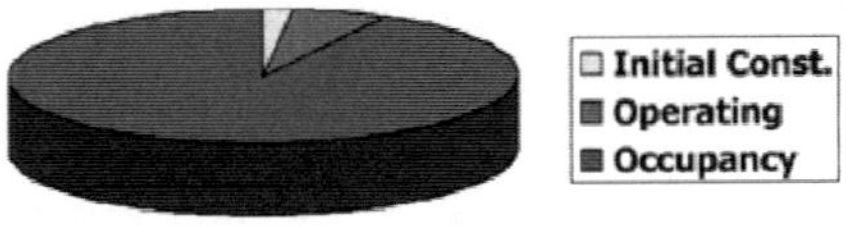

Cost of operation over 30 years for an office building

People costs account for 92% of all costs over a 30 year period.

If natural daylighting, self-control of lighting, improved indoor air quality, and all the other green building factors improved productivity by just 1% that would save the company >$1M/year for a large office complex.

Plus, if people like the building and control over their space it can give companies a recruiting advantage for top talent.

TEXAS INSTRUMENTS

Office Efficiency

- Passive solar orientation with exterior shading
- Optimized glazing (high VLT, low SHGC, low U value)
- Reflective roof
- Natural daylighting with light shelves
- High-efficiency lighting (motion and daylight sensors)
- Demand-controlled ventilation
- Attention to detail on insulation and infiltration
- Solar water heating

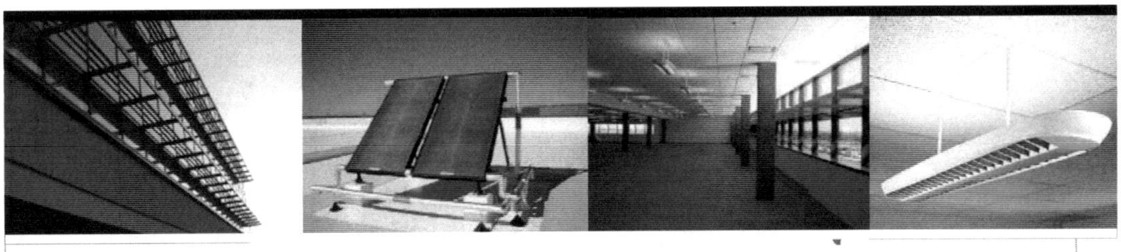

More Integration

< Areas were restored to native prairie grass to minimize irrigation and provide biodiversity.

Pond collects runoff from most of the 92 acres. 2.7 million gallon (10.2 million liters) base + 2 million gallon buffer. The pond meters runoff and settles sediment. Pond water is used for all site irrigation.

Windmill drives an air compressor to aerate the pond.

TEXAS INSTRUMENTS

Sustainability at RFAB

Native Meadow Restoration

Rain Water Reuse Pond

Reflective Roof

Dark Skies Friendly Lighting

Exterior Shades

Day lighting

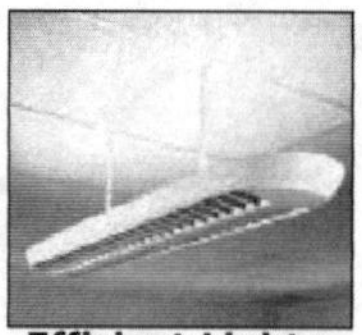

Efficient Lights

Solar Water Heating

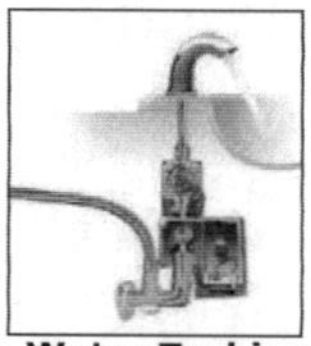

Water Turbine Powered Faucet

Bicycle Parking

Efficient cooling system with waste heat recovery

28

TEXAS INSTRUMENTS

RFAB Results

- We invested less than 1 percent of the project cost ($1.5M/$320M) in LEED-related items:
  - Predominately efficiency improvements that we would have considered regardless of LEED
- Overall project cost was 30 percent <u>less</u> per square foot than our previous fab just 6 miles away
- In the first year, we saved >$1 million in operating costs
- At full build out, we will save more than $4 million per year:
  - 25 percent energy reduction (have now reached 39%)
  - 40 percent water-use reduction
  - 50 percent on site emissions reduction
- LEED Gold Certified Office and Fab (first LEED Gold fab in the world)

www.ti.com/rfab

TEXAS INSTRUMENTS

TI Results

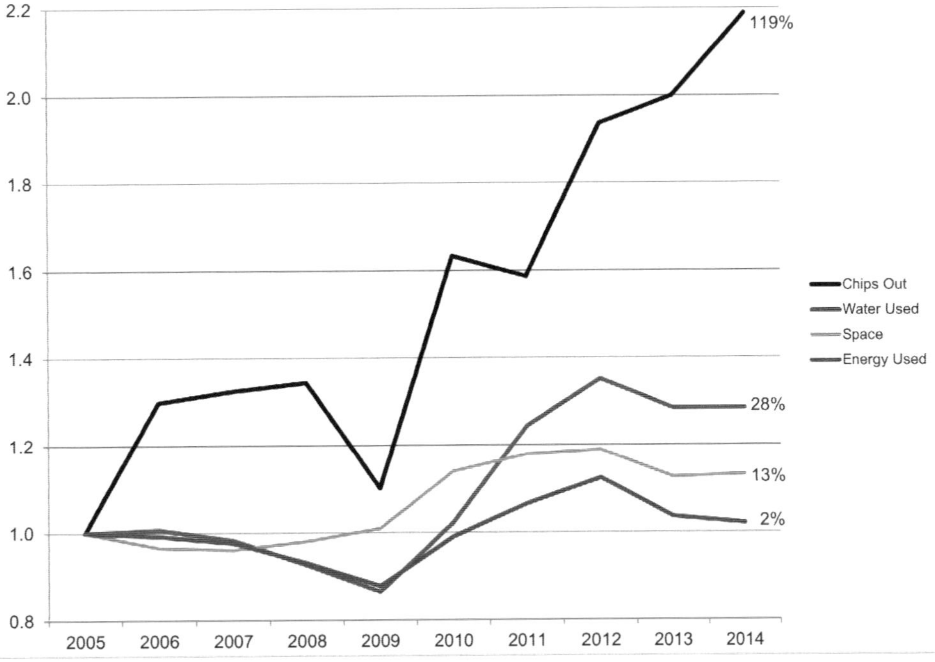

TEXAS INSTRUMENTS

Summary

- Develop the basic project requirements
- Set the sustainable goals early
- Organize for success
  - Have a broad team on board early
  - Assign a champion and integrator
  - Have a design session to exchange ideas, stretch thinking, and make connections
- Let form follow function
- Iterate and optimize the design
- Have multiple team members (example: the estimator and construction contractor) on board from the start
- Make sustainability a front page priority

TEXAS INSTRUMENTS

Bonus Ideas

- Design to operational performance goals – beyond code.
- Structure the contract to award / penalize for actual building energy and water efficiency performance
- Structure management bonuses contingent on reaching sustainable goals

TEXAS INSTRUMENTS

Don't Wait . . . Integrate

$\int_{Start}^{Finish}$ Design = Better Results

TEXAS INSTRUMENTS

Thank You

- Links
  - TI Citizenship Report – www.ti.com/ccr
  - TI RFAB – www.ti.com/rfab

  - Westbrook House – www.enerjazz.com/house
  - Paul's Prius – www.enerjazz.com/prius

Paul Westbrook
Sustainable Development Manager, Senior Member of the Technical Staff
Texas Instruments Facilities
Senior Fellow – US State Department Energy & Climate Partnership of the Americas (ECPA)
LEED AP

TEXAS INSTRUMENTS

402449 Leadership and Management of Chemical Mega Projects

Abstract:

A complex mega-project has many moving pieces. No matter how big or small a project, safety must be first priority. On a chemical mega-project it is even more critical. A mega project challenges the usual team structures. No one can do everything. The shear "force of will" approach cannot save the day as prior experience may have allowed. Structure, program and organization must be defined and practiced. Finding common ground in seemingly standard methods of communication like meeting notes, P&IDs, and mass and energy balances must be considered. Education and communication moving people form their comfortable past experience to the application of the similar methods and techniques without causing great strife among the project must be addressed. Simplification and routine is the key. Setting a regular pace structures the program to manage the multiple conflicting time and tasks. Interdependencies with flexibility are possible when designing to a common point. The definition of technology and project metrics allow all to see how to keep score. Expect what you inspect. Managing the multiple trades and crafts on a mega project construction requires the leaders to set the example and earn the respect of the crew. The program cannot become the reason why things cannot be done. Simplification has to extend into the mega project document control. Less pieces and parts enables compliance and alleviates confusion when things hit the fan. Inspection, quality assurance and quality control must be visible, active and aggressive. Millions of non-obvious details cannot be inspected, however, establishing a culture of excellence throughout the program establishes the expectations for the workmanship of all trades and crafts. The pinnacle of the effort is the pride and excitement when commissioning multiple interdependent unit operations and delivering the financial performance for the owner knowing that the program leadership made it possible.

Copyright © American Institute of Chemical Engineers. All rights reserved.

402453 Front End Loading – Methods and Examples

Abstract:

The cost of doing it right the first time is much less than repeating and spending good money to fix bad. Right the first time requires planning and collaborative understanding. The Front End Loading (FEL) workflow has become the adopted standard to ensure clear definition and active communication of the expectations and completeness of the various details that make up scope for projects. Scope definition establishes the investment decisions for the effort. A challenge in discussions of various project ideas is the experience by participants are preconditioned one there understanding of value. For example, to a business analyst, the cost of a vessel might be simplified to the cost of the materials, whereas to a seasoned project manager the multiplicative factor of 4 or 5 must be considered for all the ancillary items required to take that simple vessel and make it fully safe the functional at the commercial scale. The Front End Loading method ensures that enough detail is discussed so that expectations can be built upon common ground. The tension usually comes when the reality of the cost and effort it takes to develop a meaningful FEL Report versus just tacking the label onto a sub-par FEL effort. Any FEL effort should be a progressive build of detail as the different quality levels of the FEL effort are realized. Beginning with the traditional ideation Block Flow Diagram (BFDs), a Work Breakdown Structure (WBS) emerges. As equipment and process chemistry is developed, they are documented in the Process Flow Diagram (PFDs). At some point, the simple excel spreadsheet Mass and Energy Balance (MEB) gives way to one of the four software platforms that supports a working model of the MEB that can be manipulated and scenarios analyzed as bench learnings are incorporated for model validation. The functional descriptions or process narrative allow the first round of Piping and Instrumentation Diagrams (P&IDs) and Preliminary Hazard Analysis. Major issues can be identified for further detail analysis. General arrangements and a space allocation 3D model gives the civil and structural colleagues a chance to optimize the organization of the available space. The team continues to work with various vendors so that the major equipment specifications and the commodity equipment common bases can be developed for strategic sourcing and maintenance support simplification. Costs can be assigned as the preliminary automation and control systems master plan helps to drive the balance between manual operations and the needed crewing and the automated functions that are commonplace in a chemical plant. The FEL progresses and the project execution strategy begins to take shape as the Long Lead Procurement Item List, Design, Construction, and Commissioning Summary Level Effort Schedule, Capital Cost Estimate, Operating Cost Estimate, Construction Strategy and Permit Strategy are made available for a progressive gated approval process to insure expectations are understood and risks are mitigated to insure project success.

Copyright © American Institute of Chemical Engineers. All rights reserved.

402454 The Challenge of Skilled Labor on Chemical Mega-Projects

Abstract:

Know a good welder? How about productive electrician? So much work has become available in the chemical industry coupled with the long dry spell of investment in the chemical infrastructure of the United States that many open skilled labor positions go begging for candidates. Trade schools are not as productive as once they were, and the years of experience is embedded in a few older skilled laborers that have made it thought the slower periods of the past 20 years by being deeply embedded in the crews of the work that was available. Developing skilled labor for chemical plants is challenging because the investment that must be made may never pan out as the market forces that drive the variability in the project loads do not account for the ability of the labor pools to carry the various trades between jobs. Many times projects are dependent upon a few experienced skilled individuals that leverage themselves to support the new labor pool members that show dedication and propensity to develop the skills needed to become proficient at some of the more difficult skills needed on chemical projects. Mitigating the pipefitter shortage is the use of easier to join materials or the increased use of prefabricated spools versus the traditional stick build in the field approach. The balance of pressure of being able to complete a job with the flexibility needed due to fast paced market conditions versus the previous simplicity of an on stream strategic system build. Attracting and retaining the skilled labor pool can be facilitated by offering non-traditional schedules to travelers who are always a part of various mega-project efforts. Man camps have been replaced by the overrun of various local hotels, or the crowding in at the local trailer parks the various 5th wheels that make up housing for the crew. It is not uncommon to have a project win out preference over another by the opportunity for hours worked to be increased. The bane of a project management team is the lower productivity of the crew during difficult environmental conditions or when the balance between optimum productivity and overwork for the desire of being on the clock and racking up the hours allows fatigue to set in and effect the productivity of the crew. Men, materials and machines, maintaining the optimum balance is the key to a mega project construction execution and many of the current even big name companies are lacking in people experienced in leading the scale of projects on the drawing boards thanks to the resurgence of the American chemical industry.

Copyright © American Institute of Chemical Engineers. All rights reserved.

Selecting Individuals for One of the More Critical Jobs – Control Operator

Doug Rosenthal, Ph.D.
Job Performance Systems, Inc.

JPS
JOB PERFORMANCE SYSTEMS

Agenda

- What is it that makes the Control Operator job unique ?
- What are some of the latest scientific findings about this job ?
- How should you select your next Control Operator ?

Short Video - Refinery Control Operator

What Are Some Things That You Observed About This Job ?

What Is It That Makes This Job Unique ?

What Makes the Control Operator Job Unique

- High Information Processing Demands

Job Performance Systems, Inc. 5

What Makes the Control Operator Job Unique

- Visual Displays

Job Performance Systems, Inc. 6

What Makes the Control Operator Job Unique

- Interface Separates the Person From the Process

Job Performance Systems, Inc. 7

What Makes the Control Operator Job Unique

- Long Training Times

Job Performance Systems, Inc. 8

What Makes the Control Operator Job Unique

- **Grave Consequences of Error**

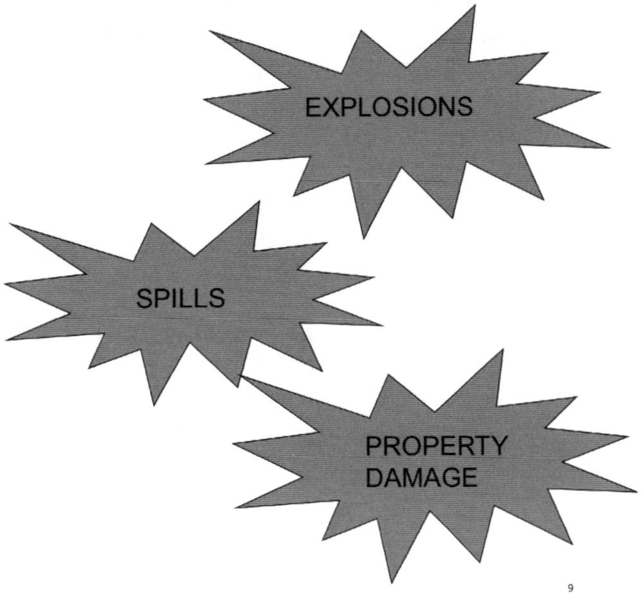

What Makes the Control Operator Job Unique

- **Stressful**

Our Scientific Findings

- **Based on Over 40 yrs of Research and Hands-On Operations Experience**
- **Job Analysis Studies at Dozens of Companies**
- **Refining, Chemical, Pipeline, Oil & Gas, Utilities**

Job Performance Systems, Inc.

11

Our Findings: Key Control Operator Cognitive Abilities

- Selective Attention - **focusing during busy and slow periods**
- Problem Sensitivity - **determining when something is likely to go wrong**
- Time Sharing (i.e., Multi-tasking) - **shifting between several things that need attention**
- Deductive Reasoning - **figuring out the cause of a problem**
- Resistance to Premature Judgment - **avoiding "over-operating" the system**
- Response Orientation - **quickly choosing among several actions**
- Visualization – **forming a mental image of the system; "seeing the big picture"**
- Speed of Closure - **making sense out of large amounts of information quickly**

Job Performance Systems, Inc.

12

The Special Challenge of Multi-Tasking

- **In general, multi-tasking is a poor way to perform work**
- **The ability to multi-task varies greatly across individuals**
- **Those who say they are good at multi-tasking may not be**
- **The trend has been to make control operators do more multi-tasking**

Job Performance Systems, Inc. 13

Our Findings: Not Everyone is Suited for This Job

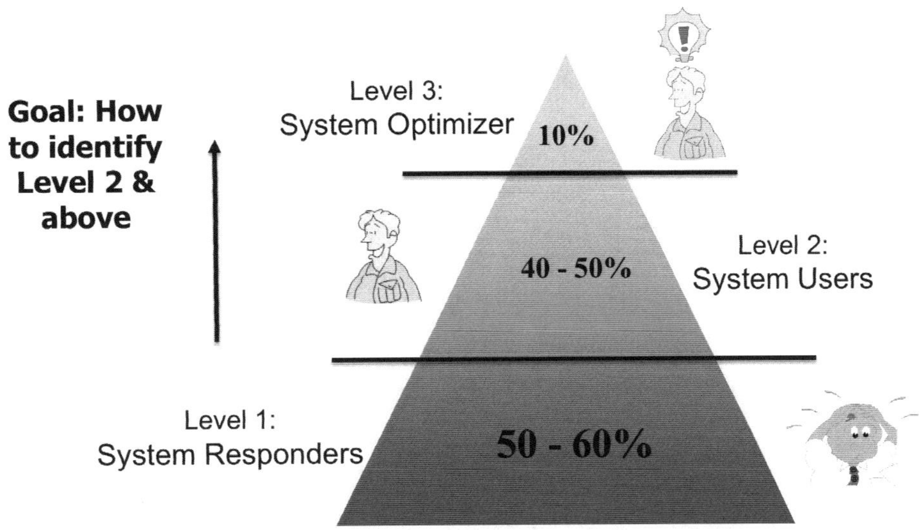

Job Performance Systems, Inc. 14

What is COBRA?

<u>C</u>onsole <u>O</u>perator <u>B</u>asic <u>R</u>equirements <u>A</u>ssessment

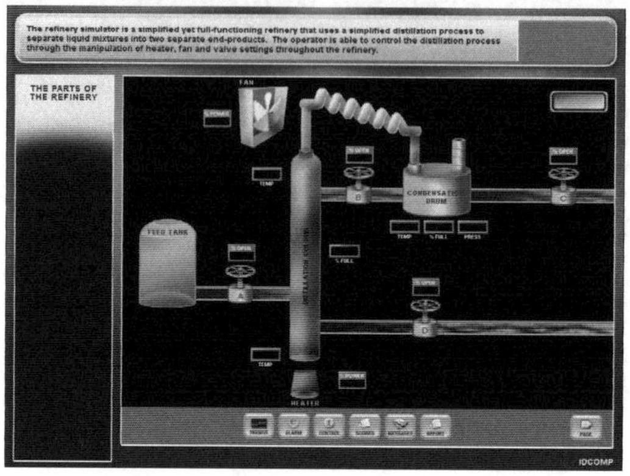

Job Performance Systems, Inc.

15

What is COBRA?

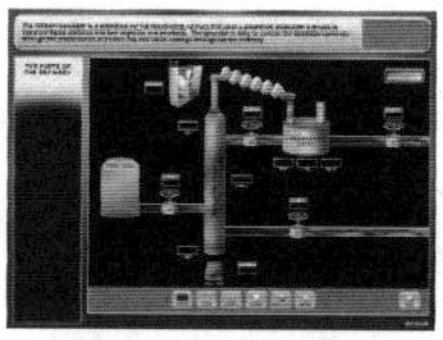

- **A simulation-based tool/test that measures aptitude and trainability of individuals**
- **Measures abilities <u>required</u> of console operators:**

  concentrate over long periods of time; concentrate in the face of distractions; multi-task; quickly detect anomalies; remain calm and focused in emergencies; and be proficient at reasoning and problem solving

- **Provides objective, computer-scored assessment**

COBRA=<u>C</u>onsole <u>O</u>perator <u>B</u>asic <u>R</u>equirements <u>A</u>ssessment

Job Performance Systems, Inc.

16

Why COBRA as a pre-screen?

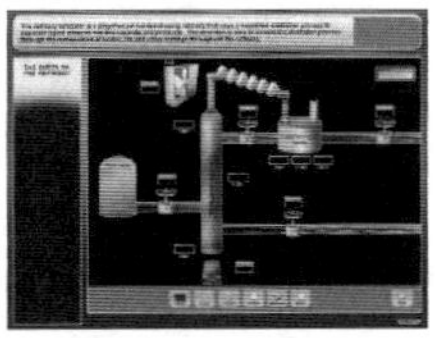

- Functions as a trainability test
- Provides a safe environment for an individual to experience the control operator role
- Allows candidate to discover if they think this work is a good fit for them
- Empirically validated

COBRA=Console Operator Basic Requirements Assessment

How Does COBRA Work?

- Anyone can take it, no experience is necessary to do well on the test
- The tool teaches you everything you need to know during the session
- 3 phases during COBRA session:
  - Phase 1 - Computer Based Training (~2 hours)
  - Phase 2 - 8 Practice Scenarios (1 hour)
  - Phase 3 - 4 Test Scenarios (1 hour)

Sample COBRA Scoring Report

- **Comments on this particular report:**
  - Scores range from 45 to 95
  - Five of the nine (55%) who were tested failed (score < 70)

Applicant	Session Date/Time	Total Score	Pass/Fail
Madison, P.	Test on December 20, 2007 8:00 AM	95.27	Pass
Smith, M.	Test on December 20, 2007 8:00 AM	86.14	Pass
Kirkwood, P.	Test on December 20, 2007 8:00 AM	75.78	Pass
Wright, C.	Test on December 20, 2007 8:00 AM	72.62	Pass
Breski, R.	Test on December 20, 2007 8:00 AM	67.58	Fail
Long, L.	Test on December 20, 2007 8:00 AM	63.99	Fail
Parsons, M.	Test on December 20, 2007 8:00 AM	62.39	Fail
Alpert, B.	Test on December 20, 2007 8:00 AM	52.28	Fail
Lane, K.	Test on December 20, 2007 8:00 AM	45.10	Fail

Job Performance Systems, Inc. 19

Our Findings: COBRA Fills In Many Gaps in the Hiring Process

Traditional hiring methods measure some "Key" abilities:

Key Abilities	Job Application	Paper & Pencil Tests	Interviews	COBRA
1. Selective Attention				★
2. Problem Sensitivity				★
3. Time Sharing				★
4. Deductive Reasoning		★		★
5. Oral Expression			★	
6. Oral Comprehension			★	
7. Resistance to Premature Judgment				★
8. Speech Recognition				
9. Response Orientation				★
10. Visualization		★		★
11. Speed of Closure				★

Job Performance Systems, Inc. 20

Our Findings: Big Payoffs to a Good Selection System

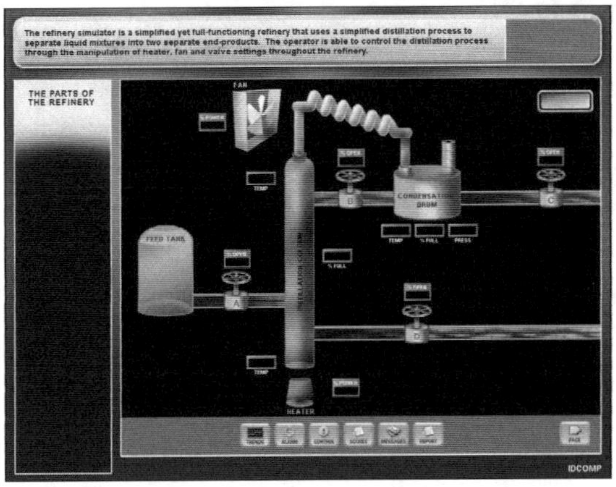

- **Highly competent operators**
- **Less training time**
- **Less attrition**
- **Fewer incidents**
- **More efficient operation**

COBRA will ensure you get high quality trainees into your workforce

Areas for Future Research

- **Structured Interview**
  - **Focus upon dependability, communication skills, leadership potential**
- **Test of ability to tolerate shift work**
  - **Realistic job preview?**
  - **Simulation?**
  - **Other?**

Summary

- **The Control Operator job is unique**
- **Our scientific findings indicate that not everyone is suited for this type of work**
- **You should use COBRA to help assess key abilities and identify who is trainable for the job**

Job Performance Systems, Inc.

23

Questions/Comments?

- **What sorts of skills and education are you looking for currently ?**
- **What challenges do you encounter in recruiting and assessing applicants ?**
- **What are typical problems your new workers have in learning the job ?**

Job Performance Systems, Inc.

24

Thank You !

Doug Rosenthal
President

DRosenthal@jps-usa.com
Phone: 703 799 3652

www.jps-usa.com
www.get-oe.com
www.get-cobra.com

JPS
Job Performance Systems

Doug Rosenthal

- **Ph.D. Industrial/Organizational Psychology**
- **Founded Job Performance Systems in 1997 to promote Safety and Operational Excellence by helping clients implement state-of-the-art Operator Selection and Training**
- **Designed the COBRA assessment test that is used at over 150 locations worldwide**

402422 Organizing the Experimental Order

Abstract:

A new process or product is an exciting undertaking. Exploring the routes of reaction and the details of separating and purifying the final product is an adventure. Safety considerations from lab to pilot scale and then finally to commercial operations must be first on everyone's mind. The experimental pathway is cloudy at first but as experiments are completed the data begins to the point the way. An organized and structured experimental approach that leaves enough room for the exploration of new ideas must be balanced. The complex Interdependencies of managing multiple experimental runs begins by taking some of the routine complexity out of the equation. Analytical measurement enable the creation of the various correlations. Results are confounded unless the analytical measurement methods can be confidently repeated and relied upon. One of the best and simply structured, but wickedly complex, content tools is the mass and energy balance (MEB). Keeping the run plans and results organized is critical so the correlations can be referenced to support investment, develop the project and defend intellectual property claims.

Copyright © American Institute of Chemical Engineers. All rights reserved.

402423 Scale up Techniques and Traps from Small Scale Systems

Abstract:

The objective of commercial scale industrial plant operation is to efficiently make one or more products over a long period of time in order to create cash flow for investors. When scaling up a technology the objective of a pilot plant is to obtain data over a short period of time that can be used to minimize technical and financial risk of subsequent larger scale designs of the same process. The objective of one is to accumulate money and that of the other is to accumulate knowledge. The logical conclusion is that different design and fabrication methods must be applied to each of the scales in order to achieve their respective objectives. Following a discussion of the differences in objectives between pilot plant and full scale design the presenter will outline practical techniques for minimizing technical and financial risk, and highlight important traps to be avoided when designing your pilot plant.

Copyright © American Institute of Chemical Engineers. All rights reserved.

White Paper

Collecting, Organizing, and Presenting Data from Small Scale Systems

Revision 0

April 15, 2015

Document ID: WP_AICHe-2015-001

Author: Mark T. Fisher
AS-EET, BS-EET, ISA-CAP
Director
Thermo Systems, L.L.C.
84 Twin Rivers Drive, East Windsor, NJ 08520
Phone: (609) 371-3300 – Fax: (609) 371-3400
www.thermosystems.com

Table of Contents

1. Abstract

Many operations and systems are awash in data. The challenge faced by scientists, engineers, and executives, is to formulate this data into actionable information. These different stakeholders each have unique information needs and today's pilot plant engineer's need to deploy information tools to meet those needs. The deployment of a successful data collection system requires a clearly defined set of user requirements followed by a solid automation plan. Critical parameters, correlations, and key performance indicators need to be defined upfront. Following these guidelines, development teams can quickly and cost effectively turn seas of data into useful information that enables them to move the pilot plant effort forward.

The pilot plant is the step between the laboratory and the full scale commercial plant and the data requirements of each type of system are unique. A scientist working on a bench in a laboratory has different data requirements than an engineer who is designing a pilot plant. Similarly, an operator working in a commercial plant cares more about yield, throughput, and cost per unit produced, than about mass balances and the scientific method which are the realms of the pilot plant engineer and the scientist, respectively.

Understanding the unique data requirements of each of stakeholder is fundamental to developing a robust pilot plant data plan. This white paper presents concepts to consider when developing a process data system for a typical pilot plant. It does not focus on details such as what process variables to monitor or how to monitor them, but instead presents some best practices to consider in order to quickly and efficiently transform raw process data into actionable information.

2. Background

This section provides background information necessary to fully understand the challenges that are encountered when designing and implementing a typical pilot plant data system.

2.1. Data vs Information vs Knowledge vs Wisdom

First and foremost it must be understood that data is NOT information. Consider the standard DIKW pyramid below.

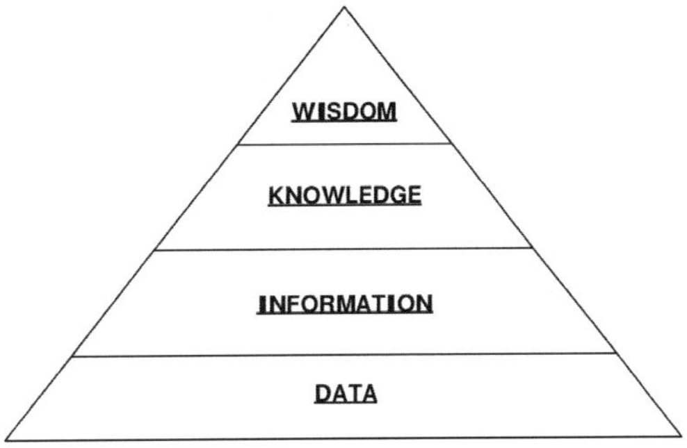

Figure 1 – DIKW Pyramid

Data are raw facts. They have no meaning in of themselves. They are neither right nor wrong. They just are.

1. The object is red. The object is round. The object weighs 4 ounces.
2. The pressure in the vessel is 125 psi.

Information uses data. When data is processed, organized, structured, and/or presented in a given context it becomes information. Information can be right or wrong, depending on the context that has been given to it.

1. The object is a fruit. This eliminates things like balls and jaw breakers but does not give enough context to the object to determine exactly what it is. It could be an apple or a tomato, both are fruits. If however, more context is given, such as the object grows on vines not trees, then it becomes fairly obvious that the object is a tomato, not an apple.
2. The pressure rating of the vessel is 150 psi. The pressure is increasing.

Knowledge uses information to come to some conclusion. When multiple pieces of similar information are used together to it becomes knowledge. Knowledge is the understanding of patterns of information.

1. Even though a tomato (the object) is a fruit, it is rarely used in fruit salad.
2. If the pressure continues to increase above the vessel's rating (150 psi) the vessel will burst.

Wisdom uses knowledge to correct, alter, or prevent a situation. It is the action that results from knowledge.

1. I won't put tomatoes in my fruit salad.
2. I will open the pressure relief valve on the vessel to relieve the pressure before it bursts

Actionable Information. The term "actionable information" is used throughout this paper to refer to *data* that has been transformed into *information* that *knowledge* can be gleamed from and *wise* actions can be applied to.

2.2. Transforming Data Into Actionable Information

So now that you understand the difference between data and actionable information, let's focus on the aspects involved in transforming raw process data into actionable information in a typical pilot plant.

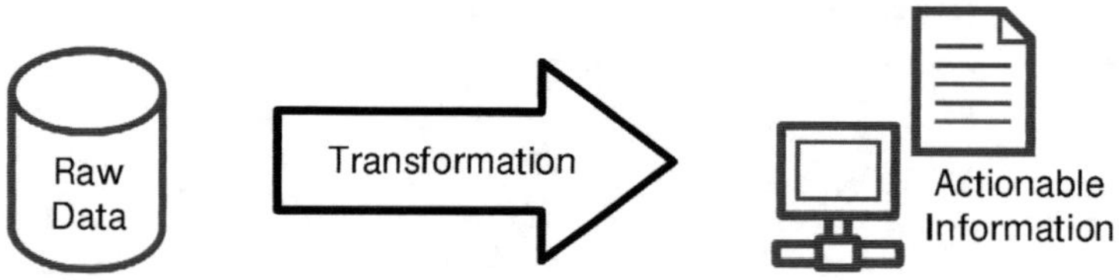

Figure 2 – Transforming Data into Information

2.2.1. Multiple Stakeholder Sets

As with many processes, sometimes the best place to start is at the end. This is true of the data transformation process in a pilot plant. It must be understood that there are several different types of people who will need access to the data from your system. These groups of people are referred to as stakeholder sets. A typical pilot plant data system may have the following stakeholder sets:

- Scientists
- Engineers
- Maintenance Personnel
- Operators
- Supervisors
- Quality Assurance Personnel
- Plant Executives
- Corporate Executives
- Board Members

Your particular pilot plant may include some or all of these stakeholder sets and maybe even some others. The important thing to remember is that each set will have very different requirements concerning how the data that is stored in your system is transformed into actionable information.

2.2.2. Divergent Data Sources

With the end in sight, now let's focus on the beginning. Most pilot plants will have more than one source of data. The figure that follows illustrates a typical pilot plant with four (4) divergent sources of data.

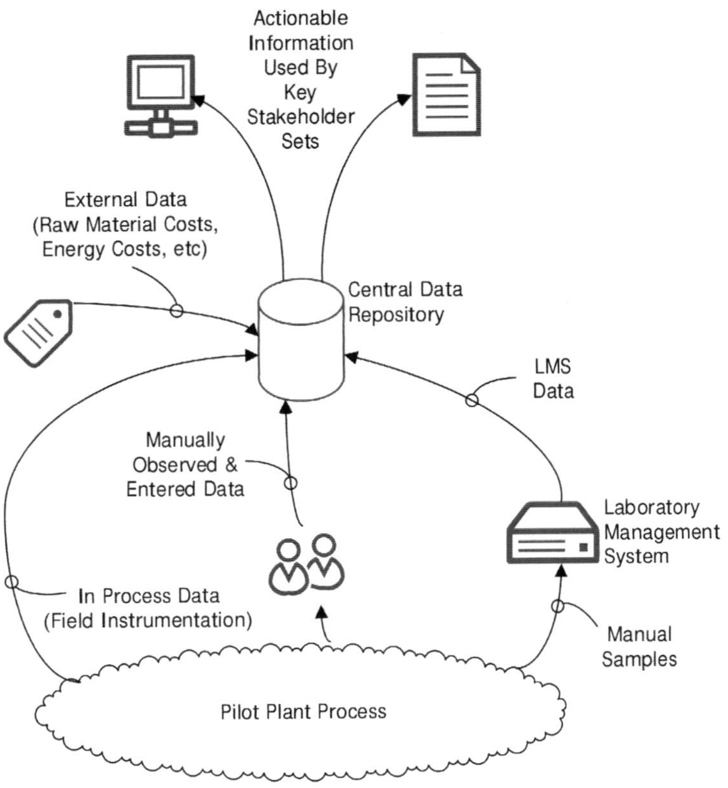

Figure 3 – Pilot Plant Data Collection Model

As illustrated in the figure, there are four (4) different data sources that provide data into the system.

1. In Process Data (Field Instrumentation)
2. Laboratory Equipment
3. Manually Observed and Entered Data
4. External Sources

Your particular pilot plant may have some or all of these data sources and maybe even some others. The important thing to remember is that there are often many different ways that data can be delivered to your system, and its design must be flexible enough to handle all of them.

In Process Data is data that is measured "real-time" directly at the pilot plant process itself. It is generated from field instrumentation that is installed directly in or near the process stream in order to safely and efficiently control and monitor the pilot plant process. The data from some of this instrumentation may also be used to meet the information requirements of some of the key stakeholder sets but its main purpose is process control. This instrumentation is almost always wired in to the process control system that monitors and controls the pilot plant process. Instrumentation of this type is beyond the scope of this paper unless it is also used to meet a key stakeholders information requirements. Note that some in process instrumentation may be installed for the sole purpose of meeting a key stakeholder's information requirements and have nothing to do with actual process control at all.

Laboratory Equipment. It is impossible to measure some variables "in or near process" due to the technology currently available. For these types of variables a sample must be taken manually (or automatically in some instances) and analyzed in a laboratory. The data from the laboratory must then be input into the pilot plant data system's central data repository, either manually, or preferably via some automatic electronic transfer mechanism.

Manually Observed and Entered Data. Other types of variables or conditions cannot be measured at all, even by currently available laboratory equipment. For these types of variables it may be required that an operator "tour" the plant at regular intervals and enter their observations manually either with paper and pen via log sheets, or preferably through some sort of electronic data entry means such as a tablet computer or similar device.

External Sources. The types of data that may enter your pilot plant's data system from external sources include, but are not limited to, cost of raw materials, raw material inventories, and cost of energy. This type of information may be entered into your system manually but ideally it would come in automatically in semi-real time using RSS feeds or some other related technology.

2.2.3. Data Transformation Process

Referring back to Figure 2 on Page 4, the act of transforming raw data into actionable information is a bit more complex than the figure implies. It is actually an intricate process that involves multiple steps:

1. Collection
2. Storage
3. Processing and Manipulation
4. Retrieval
5. Presentation

Collection involves the communication and aggregation of the raw data from the multiple data sources described in the previous section at a central location.

Storage is the act of placing the aggregated data onto a media, such as a hard drive, so that it can be retrieved at a later date.

Processing and Manipulation involves qualifying the raw data against validity and accuracy rules and in some cases flagging it as out of range or otherwise changing it so that it can be presented properly in subsequent steps in the transformation process.

Retrieval is the act of transferring the qualified data from storage and delivering it to the next step in the transformation process, the presentation step.

Presentation is the last step in the data transformation process and involves displaying the qualified data that was retrieved in a format that gives it context, thus turning it into actionable information.

3. Challenges

The sections that follow discuss some of the challenges that you are likely to run in to when developing and implementing a data system for a typical pilot plant.

3.1. Multiple Stakeholder Sets

As mentioned previously, there are many stakeholder sets who will need access to the raw data that your pilot plant data system will collect. Furthermore, each set will need that data presented differently in order to give it the context needed to turn it into actionable information that can be used to fulfill their job function.

The challenge here is to gather all of the requirements from representatives of each stakeholder set and to fully understand and communicate what your understanding of what those requirements are. You must be sensitive to the fact that some of these key stakeholders are plant and even corporate level executives. Keep in mind that in larger organizations, people at this level of management may have little or even no knowledge of what you are doing in the pilot plant. None the less, if they are identified as a member of a key stakeholder set, their input is critical to your system's success.

3.2. Divergent Data Sources

The fact that data can enter your pilot plant data system through various avenues presents its own set of challenges.

Each data source is likely to use a different protocol or method for getting data out of it and in to your system. You need to identify what these protocols/methods are and you need to ensure that your system is capable of handling them. Error checking, fault tolerance, and redundancy of the data collection mechanisms needs to be considered at every level.

A way of linking data from all of the pertinent data sources must be provided, so that when the data is presented to the stakeholders it is grouped together regardless of where it came from or how it entered your system. The source of the data needs to be transparent to the stakeholder sets who will be using your system.

Manual data entry present its own set of challenges and should be avoided whenever possible, but the reality is that there will likely always be a need for manual data entry. Your system needs to minimize the chances of manual entry and transcription errors, eliminate them altogether if possible, and to at least flag and/or clamp them to valid limits whenever possible.

Data from external sources needs to be considered very carefully from a cyber-security point of view. The last thing you want to do is to leave your pilot plant vulnerable to outside threats from hackers, malware, your competition, or even other nation states. Loss of intellectual property, unauthorized data manipulation, loss of process control, regulatory violations, loss of public and/or regulatory confidence, equipment damage, and even personnel or public injury or death are only some of the consequences that can occur if a well thought out cyber security plan is not implemented.

3.3. Data Transformation Process

Obviously, the act of transforming raw data into actionable information is in of itself a challenge. The items that follow list some of the considerations that you should take into account at each step of the data transformation process. This is by means an exhaustive list. The items that are presented are meant to give you an idea of the types of things to consider during each step. You will undoubtedly encounter others when designing and deploying your own pilot plant data systems.

1. Collection
    a. Determination of collection frequency for a data point or groups of points
    b. Whether to use exception or polling based collection
    c. If and how data should be compressed in order save storage space and increase retrieval time
    d. How to ensure collection fault tolerance
    e. Collection redundancy
    f. How will manual date entry be performed and collected
        i. Consider tablet PCs or some other portable computing device that your operators can carry around as they tour the plant
        ii. Portable computing devices should be able to be tied directly in to the central data repository via wireless technology
        iii. Consider bar codes and/or RFID tags and scanners to minimize data entry and transcription errors
        iv. Clamp manual entry points between acceptable limits
        v. Manual entry software should be able to be configured with "virtual tours" that walk the operator through his routine tours of the plant

2. Storage
    a. How big do my storage devices need to be
    b. How long do I need to retain my data for
    c. Data archival, backup, and restoration methods

3. Processing and Manipulation
    a. How to qualify a data point or group of points
    b. Consider using the following factors in your data qualification schema
        i. The detectable range of the instrument that generated the data
        ii. Instrument range limits
        iii. Confidence levels of the operators when manual samples are taken and laboratory analysis is conducted
        iv. Confidence levels of lab procedures
        v. What to do with raw data once it is qualified and found to be of bad quality
        vi. Typically you'll want to keep raw data so that it can be requalified at a later if the qualification parameters change

4. Retrieval
    a. Retrieval methods should be easy to use by non-technical personnel
        i. Whenever possible programming languages like SQL should be avoided
        ii. Instead use of pull down boxes, selection lists, and other GUI objects
    b. Consider modeling your pilot plant's physical assets into a framework that is easy to understand and uses "human" terminology to describe it
        i. Example: Use "Batch Tank 100 Discharge Flow" as opposed to "TK100-FIT01"
        ii. The ISA-S88 standard is excellent for modeling most types of systems and grouping them into
            - Process Cells
            - Units
            - Equipment Modules
            - Control Modules
        iii. The hierarchy of the asset framework should utilize reusable templates that can be combined and build on each other
    c. Automatic vs manual retrieval
    d. Selection of retrieval methodology for:
        i. Canned reports
        ii. Ad-hoc, one-off reports that easy to configure even by non-technical staff
5. Presentation
    a. How best to present data in order to give it the context needed to transform it into information and that is usable by all of the stakeholder sets
    b. How to present the data in both printed and online format
    c. How to automatically and manually generate reports in both online and printed formats
    d. For online presentations
        i. Where does the information needed to be presented to
            - Computers on the same network as the pilot plant's data system
            - Computers on the owner's corporate network
            - Computers on the internet
            - Emails, smartphones, etc.

4. Recommendations

This chapter first presents a project style, step by step approach to use to design and deploy a robust pilot plant data system that if implemented properly, will help to promote a fast and cost effective migration to full-scale production. It then goes on to address each of the challenges identified in the previous chapter and provides detailed recommendations on how to meet those challenges.

4.1. Project within a Project Approach

First and foremost, it must be understood that the design and deployment of a pilot plant data system is a project all unto itself. It may be part of the overall pilot plant project, but it should be treated as a separate project all together. The disciplines and focus of the personnel who design and deploy these types of systems are for the most part, different from those who design and deploy the physical aspects of a pilot plant.

Treating the pilot plant data system as a separate project and employing good project management methodologies will ensure its success. A dedicated project manager should be assigned to oversee the whole project. If the owner does not have in-house expertise in this field, a qualified and experienced Systems Integration (SI) Firm should be contracted to manage, design, and deploy the effort.

- Project schedules should be published and administered, milestones should be tracked.
- Resources should be allocated and managed.
- Key stakeholder requirements should be gathered, documented and approved.
- Specifications (both hardware and software) that meet those requirements should be authored and adhered to.
- Testing protocols should be executed, both at the SI's factory and again at the site once the rest of the pilot plant is mechanically and electrically ready.
- Project status meetings should be held at regularly scheduled intervals.
- Budgets should be tracked.

4.2. Systematic/Multistep Approach

The following steps should be implemented in the order indicated to ensure a smooth and efficient pilot plant data system project implementation.

Step 1. Identify every stakeholder set who could potentially require access to the pilot plant data system. Assign a single representative and an alternate (if possible) from each set who will champion the requirements of the rest of their sets' members throughout the process. Meet and interview each of those representatives to determine what their sets' Key Performance Indicators (KPI) are and how they want them presented.

Step 2. Identify the instrumentation, equipment, and data sources that are necessary to calculate/derive the KPIs identified in step 1.

Step 3. Generate a Data Requirements Specification (DRS) that lists each stakeholder sets' KPIs as well as the instrumentation, equipment, and/or data sources that are required to calculate/derive them. This is not a detailed specification, it is a requirements specification. Only requirements should be contained in this document. Detailed equipment and software specifications follow later in this process. The DRS should be reviewed and approved by all stakeholder set representatives before moving to the next step. If requirements change later in the process after the DRS is approved, then the DRS should be revised and re-approved accordingly. It should be a living document that exists throughout the data system's life cycle.

Step 4. Generate a Detailed Data System Design Specification (DDSDS) that specifies exactly what instrumentation, equipment, communication infrastructure, hardware, and software are necessary to meet the requirements outlined in the DRS. The DDSDS should be reviewed and approved by all stakeholder set representatives before moving to the next step. If requirements and/or specifications change later in the process after the DDSDS is approved, then the DDSDS should be revised and approved accordingly. It should be a living document that exists throughout the data system's life cycle.

It is important to note that the pilot plant's process control system and its data system are typically tightly integrated together and often share a lot of the same infrastructure equipment and software. It is therefore acceptable and often preferable to include the information that would normally be contained in the DRS and DDSDS in the process control system's User/Functional Requirements Specification (U/FRS) and Detailed Design Specification (DDS), respectively.

A special note on "in or near process" instrumentation is also required here. Most of the time, an "in or near process" instrument that is required to calculate or derive a KPI will already exist in the pilot plant's process control system itself. It is not the intent of this paper to imply that separate/duplicate instruments should be provided for the process control and data systems. Whenever possible, the same instrument should be used for both systems unless some other requirement such as safety, redundancy, or maintenance warrants duplication. In situations where an instrument that is required for KPI purposes already exists for process control purposes, the DDSDS should somehow flag it as such. These types of instruments will already be specified in documents generated for the process control system. The DDSDS should not re-specify these types of instruments but should instead provide a reference pointing the reader to the process control system document where those specifications can be found. Only those instruments that are required solely for data system purposes should be specified in the DDSDS.

Step 5. Procure, implement, and deploy the pilot plant data system, adhering to all of the specifications contained in the DDSDS. Prototypes of every report, either in printed and/or online format, should be provided to the key stakeholder representatives for approval prior to final implementation.

Step 6. Generate testing protocols that ensure that all of the requirements laid out in the DRS are met and that the system has been deployed according to all specifications. Preferably the data system will be staged in a laboratory setting before it is actually installed at the pilot plant site on the pilot plant's field equipment. If a System's Integrator (SI) is contracted to deploy the data system this will typically take place at their factory, but it may also be conducted at the owner's facility or some other location at the discretion of the owner. In situations like this two (2) separate testing protocols will be created. The first, referred to as the Factory Acceptance Testing (FAT) protocol will be executed prior to installation at the pilot plant site at the SI's factory or other agreed to location. Sometimes the FAT protocol is broken down into two (2) separate protocols: the Hardware Factory Acceptance Testing (HFAT) protocol, and the Software Factory Acceptance Testing (SFAT) protocol. The last testing protocol will be executed once the system is deployed to site and is referred to as the Site Acceptance Testing (SAT) protocol. All testing protocols should be reviewed and approved by all key stakeholder set representatives before they are executed.

Step 6A. Execute the Factory Acceptance Testing (FAT) protocol(s) using simulators and/or simulation logic to exercise the system to ensure proper operation. Any and all stakeholder representatives should be invited to witness the factory testing. Do not proceed to the next step in the process until all factory testing has been successfully completed and the key stakeholder representatives approve the final executed protocol document(s).

Step 6B. Execute the Site Acceptance Testing (SAT) protocol on the actual pilot plant process equipment at the site to ensure proper installation and operation. Any and all stakeholder representatives should be invited to witness the site testing. Once the site testing has been successfully completed and the key stakeholder representatives approve the final executed protocol document the system can be turned over to the owner for training and operation.

Step 7. Conduct training for all of the key stakeholder representatives and other personnel identified by the owner on the operation, maintenance, and troubleshooting of the system. Other training for the owner's engineers, service, and IT personnel may also be required if they are going to maintain and make changes to the system in the future.

4.3. Detailed Recommendations Addressing Challenges

The sections that follow make detailed recommendations to the challenges that were identified in Chapter 3 of this document. Please keep in mind that the sections that follow are only recommendations to give you an idea of what can be done. It is not the intent of this paper to suggest that they must be done. Some of these recommendations can be rather costly and the may be prohibitive for your particular application. Other recommendations may not apply to your application at all.

4.3.1. Multiple Stakeholder Sets

In order to gather all of the requirements from all of the key stakeholder sets you obviously need identify who they are. This is typically done by asking the project manager: "What departments in your organization are going to need to access the data from the pilot plant and who can I speak to find out what their requirements are going to be"? Often times the first person you speak to from each department will not be the person who will ultimately be responsible for providing you the information you need. However, you must be persistent. Failure at this stage will most likely result in failure of the whole process. It is imperative that you get the right people on board as early on in the process as possible.

Eventually you will find the right person who has the information you need and the authority to speak for the entire stakeholder set. If at all possible, also try to find an alternate representative from each set in case the primary representative is unavailable for extended periods of time.

It is important that you let each representative know what their responsibilities are going to be so that they know what they are getting into. This information should be presented to the representatives in writing and reviewed with them in person. At this point some representatives may decide to decline and will refer you to other people. That's OK, don't get discouraged, it's all part of the process.

This representative approach becomes more important the higher up in the owner's organizational chart you go. It is very unlikely that you are going to get plant managers or CEOs to sign on as stakeholder set reps. However, they will be using your data every day to make informed decisions that affect the entire organization, so their input to the process is critical to the success of the pilot plant. In these instances higher level managers, directors, and executives will likely assign someone to your team who will report back to them directly and will have the authority to speak for them as a stakeholder set representative. As difficult as it may be, getting buy in from people at this level in the organization is perhaps the most important part of this process.

A representative is also very important when the stakeholder set is comprised of a lot of people. Trying to get a room full of people to agree to anything is very difficult. If however, you have one or two people who have the authority to speak for the whole group the entire process is streamlined.

Once the key stakeholder set representatives are identified you will need to interview each of them personally to determine what their sets' requirements are. Ask questions like:

1. What are the Key Performance Indicators (KPIs) that your group is looking to monitor?
2. How does your group want these KPIs presented?

Some reps will come prepared and will tell you exactly what you need to know while others may need a little coaching and others still may not even know what a KPI is. For your reference:

KPI stands for Key Performance Indicator. KPIs are derived from one or more data points. By definition KPIs are information, not data. They are used to gleam knowledge from your process and to take action based on what the KPIs tell them. A detailed discussion on what makes a good KPI is beyond the scope of this document, but there are many online sources that can be referred to.

Choosing the right KPIs relies heavily upon having a good understanding of what is important to each stakeholder set, which is why the representative approach is so effective. You can guess what's important to other people, but more than likely you will guess wrong.

Be prepared when you interview your reps. Bring examples of KPIs and how they have been presented in other applications. This white paper includes several KPI presentation graphics that can be used as a starting point. There are hundreds of other ones available online and most historian vendors provide sample KPI templates that you can use.

The output from your interviews with the key stakeholder representatives will be a list of KPIs that you will document in a Data Requirements Specification (DRS). See section 4.2 earlier in this paper for more information on the DRS. You will use this information to decide what instrumentation and/or other equipment is necessary to calculate or derive each KPI, it is therefore the first step in the data system's definition and specification and must be done properly to ensure success down the road.

4.3.2. Divergent Data Sources

Section 2.2.2 introduced a typical pilot plant data collection model with four (4) divergent data source inputs:

1. In Process Data (Field Instrumentation)
2. Laboratory Equipment
3. Manually Observed and Entered Data
4. External Sources

The paragraphs that follow provide some recommendations concerning the design and implementation of the interfaces to those data sources.

In Process Data (Field Instrumentation) Interface – "In or near process" instrumentation is almost always wired in to the process control system that monitors and controls the pilot plant process itself. Typically this is either a Programmable Logic Controller (PLC) system or a Distributed Control System (DCS).

The main purpose of this type of instrumentation is to monitor and control the pilot plant process in real time. However, it is always good practice to collect data from this type of instrumentation in the pilot plant data system. There may also be some cases where "in or near process" instrumentation is installed for the sole purpose of providing important data to the data system and has nothing to do with the process's control at all. In either case the data will be communicated indirectly to the data system from the process control system not directly from the instrumentation itself. Therefore, it is imperative that the interface to your central data repository system support a communication protocol that the process control system also supports. You do not want to have to write a custom communication driver for this purpose. Most commercially available central data repository systems support many different types of process controller interfaces so this will most likely not be a problem unless you are using some sort of unusual process controller.

Since the majority of the data that your data system will collect will come through this data source, you may want to consider building in fault tolerance at the interface level where if the interface were to fail, data would be queued up locally on the process control system side of the interface until the interface is restored at which point the queued data would feed forward to the central data repository server. Another option would be to install fully redundant interfaces where if the primary interface were to fail then the backup interface would kick in automatically. Often times fault tolerant and/or redundant interfaces at the pilot plant level may be overkill and cost prohibitive, but it is definitely something to consider. Most commercially available central data repository systems support some sort of fault tolerant and/or redundant interfaces.

Laboratory Equipment Interface – Unlike the In Process Data Interface described above where data is continually communicated to the central data repository system, data from laboratory equipment will most likely come across in batches as manually taken samples are analyzed by the lab equipment. Furthermore, while some of your lab equipment may support one or more of the same communication protocols that the central data repository system supports, this is most often is not the case.

There are typically two options used to interface to laboratory equipment. The first option involves your lab equipment copying a Comma Separated Variable (CSV) file or some other type of delineated file to a specific location on a network drive when it completes its analysis of each sample. The file will contain the data from the sample. When the central data repository system sees a file at this location it will open it, parse it, and insert the parsed data into its data repository. This functionality is not available in all central data repository systems or laboratory equipment though. You will need to check to make sure.

The second option involves purchasing another type of system known as a Laboratory Information Management System (LIMS). A LIMS sits between your lab equipment and your central data repository system. It handles gathering the information from all of your lab equipment and sending it to your central data repository system. Specification of lab equipment and LIMS is beyond the scope of this paper, and not all commercially available central data repository systems support interfaces to LIMS. Again, you will need to check to make sure.

Manually Observed and Entered Data Interface – If your system needs to support manually entered data, the following paragraphs contain some best practice suggestions for its implementation.

First the central data repository system you select must obviously support manually entered data and provide the tools to create the online forms that the operators will enter their data on. Manual entered data has great potential for errors and the forms you design should make extensive use of drop down lists and other GUI interface objects that minimize the potential for data entry errors. Limits and other error checking methods should be employed to further minimize the chance of data entry errors. However, no matter how hard you try to eliminate entry errors they are still going to occur. That is why all manually entered data should be required to be reviewed by a supervisor or other second party before it is committed to the central data repository system.

If you are going to require your operators to perform regular rounds of your plant and to enter their observations as they make those rounds then the central data repository system you select and design must support interfacing to portable tablets or other portable computing devices that are carried around by operators as they make those rounds.

Some central data repository systems support software that allow "tours" to be created. These "tours" are online forms that run on the operators' tablets that walk them through their rounds. The forms prompt the operators to confirm that they have "visited" important locations during their rounds. It is highly recommended that bar code, RFID, or some other sort a label be placed at these locations. Scanners should then be connected to the operators' tablets that can read these labels to ensure that the operators actually "visit" each location during their rounds. The operator may be required to make an observation of some sort at each location and to enter their observations on the tour forms. In these cases the data entry error minimization techniques described above should be employed.

Note that if you are going to require your operators take samples from the process stream during their rounds for subsequent processing in the lab, the "tour" concept described above with labels at each sample port is a good way to capture what port the sample was taken from.

External Source Interface – If your application requires that data from external sources be supported then obviously the central data repository system you select must be able to communicate with those sources. RSS feeds are a typical mechanism to bring in data from external sources but there are many others. You will need to confirm compatibility between your central data repository system and the external data source.

The main thing you need to consider whenever you open your system up to external sources is security. You must have a well-defined and implemented cyber security plan in place to ensure that your system is locked down. Cyber security is beyond the scope of this paper.

4.3.3. Data Transformation Process

Section 3.3 presents a series of questions and things to consider during each step of the data transformation process. This list should be referred to often as you design your pilot plant data system because how you respond to those questions will affect almost every part of your design.

4.3.4. Central Data Repository Selection Criteria

The selection of the right central data repository for your application is perhaps the most important decision you will make during the design of your pilot plant data system.

Building your pilot plant data system around a central data repository is extremely important. Consider for a moment how difficult it would be to transform your data into actionable information if all of your divergent data sources stored their data in in their own, and potentially proprietary files, and all of those files were in different locations.

There are several manufacturers of central data repository systems, or as they are commonly called historian's. It is not the intent of this paper to recommend one manufacturer over another, just realize that the input side of any historian package is just as important as the output side. Don't be lured in by fancy graphics and reports that the vendor pitches during their sales call to you. Just about any commercially available historian has the ability to generate very sophisticated reports in the right developer's hands. It's what's under the hood that counts.

To stress this fact, refer to figure 3 earlier in this document. As you can see from the figure, the central data repository is where all of your divergent data sources send their data. Therefore, first and foremost, the historian you select should be capable of communicating to <u>all</u> of your data sources "right out of the box" without having to write custom communication drivers.

Next, you must realize that large amounts of raw data, even in small pilot plants, can be generated and accumulated very quickly. The historian you select should support both exception based and polling based data source communication. It should also incorporate sophisticated data compression algorithms so that when you begin to accumulate large amounts of data your queries won't take so long to execute that it renders the system unusable.

Of course the repository you select must also be capable of generating sophisticated reports, both in print and online formats, but as stated earlier this is the least of the functionalities you should be concerned with. Remember, it's under the hood that counts.

Another thing to consider when selecting a central data repository system is how critical the data is to you. If your answer is "very", you may want to consider installing redundant central data repository servers. Although at the pilot plant level this may be overkill in most applications. It is certainly important to consider when you go to a full scale production plant where the financial risk of having to throw away a batch due to lost data is much higher than it typically is in a pilot plant. But it is something to consider at the pilot plant level as well depending on the criticalness of your data. Most commercially available historian's support redundant servers in one form or another.

And finally, In order to bring home the importance of selecting the right central data repository, consider the standard DIKW Pyramid that was presented in Figure 1 at the beginning of this paper. As you can see from the figure, "Data" is at the bottom of the pyramid and thus forms its foundation. If your data (ie. your historian) is weak, the pyramid will crumble and come crashing down. Just like your pilot plant's data system will if you fail to pick the right central data repository for your application.

5. Benefits

By considering the recommendations presented in this paper you will be well on your way to developing a pilot plant data system that can quickly turn mountains of data into useful and actionable information that in turn will facilitate a quick and efficient transition to full scale production.

6. Conclusions

This paper has presented some of the challenges you will undoubtedly encounter as you design your pilot plant's data system and has proposed a number of recommendations to overcome those challenges. The challenges and recommendations presented are by no means an exhaustive list. Every pilot plant is unique and each will have its own set of unique challenges that you will have to overcome. But if you ensure that you know the requirements of the key decision makers involved and keep those requirements in mind throughout the entire process you will ultimately succeed.

Remember that the pilot plant is the bridge between the laboratory and the commercial production plant and its data system will operate long after the physical construction of the pilot plant is complete. It is the tool that senior management will use to determine if it is worthwhile to stay the course and build a full scale production facility or to abort the process altogether and move on to other endeavors.

The best designed and implemented pilot plant will ultimately fail if its data system doesn't deliver the information to the decision makers who need it, when they need it, and in the format that they need it in.

In conclusion, always remember that data is not information, and without information there can be no knowledge, and without knowledge there can be no wisdom to make change and move forward. Data must be transformed into information and in order to do that you must first and foremost understand the requirements of the people who need the information and who have the authority and ambition to turn it into action.

7. Sample KPI Graphics and Reports

The following pages contain some sample KPI graphics and reports for reference.

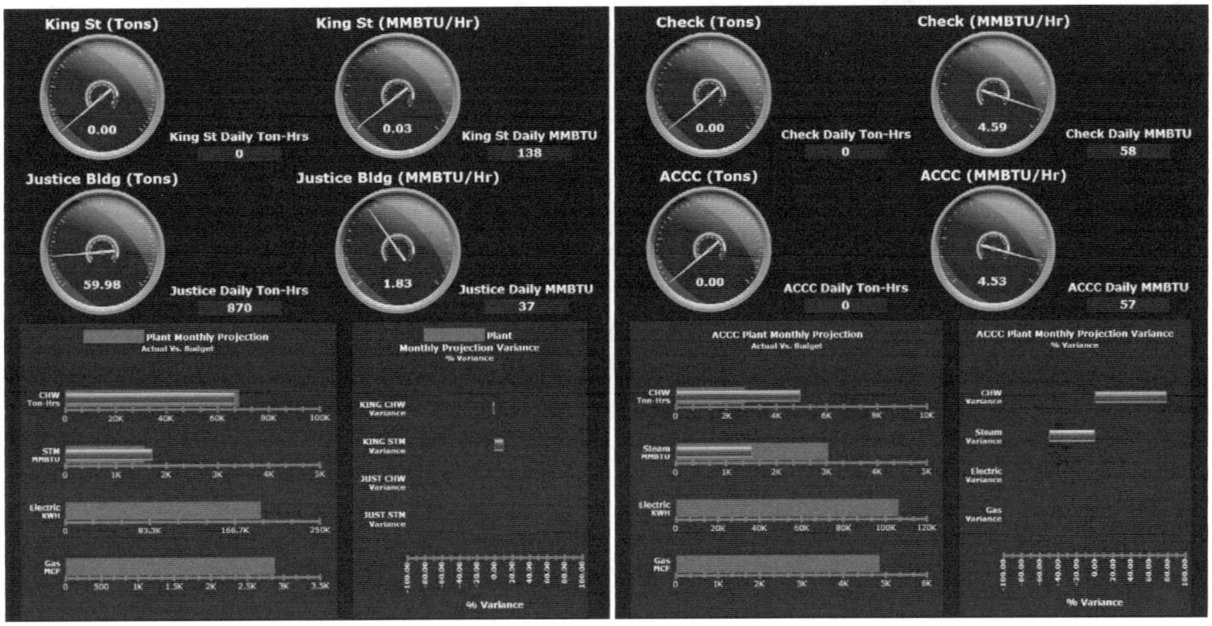

Figure 4 – Sample KPI Graphic 1

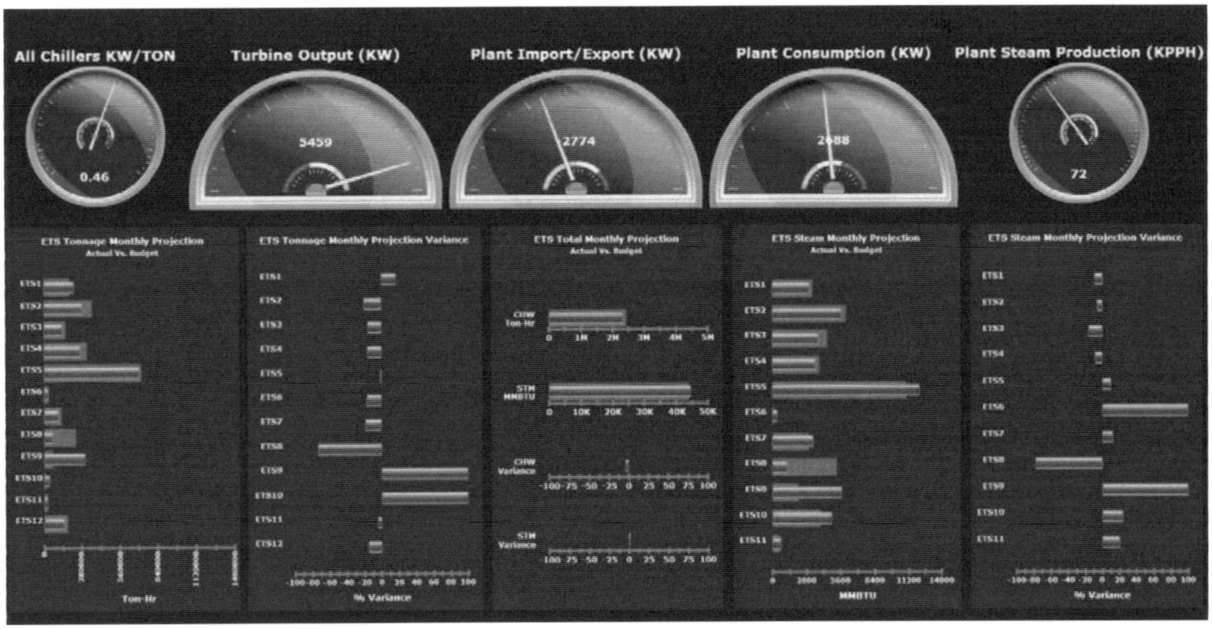

Figure 5 – Sample KPI Graphic 2

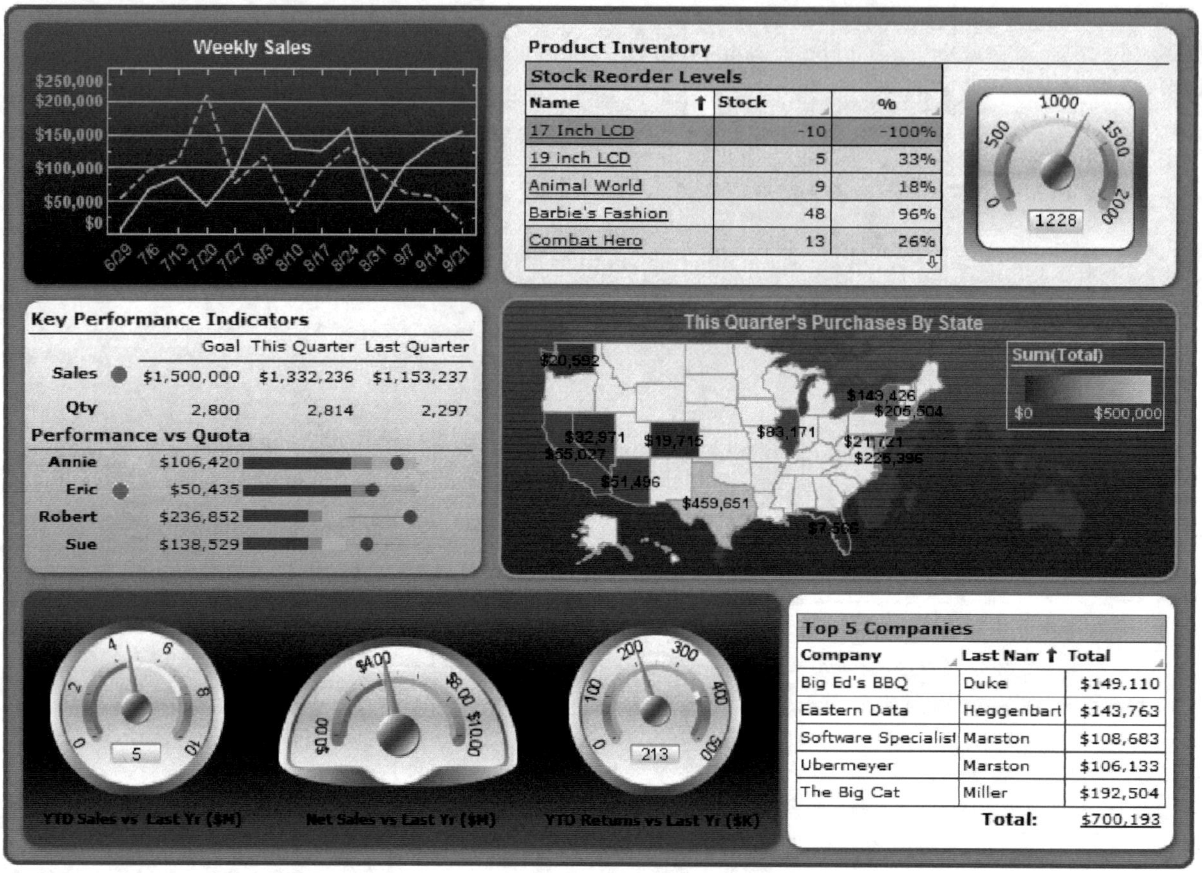

Figure 6 – Sample KPI Graphic 3

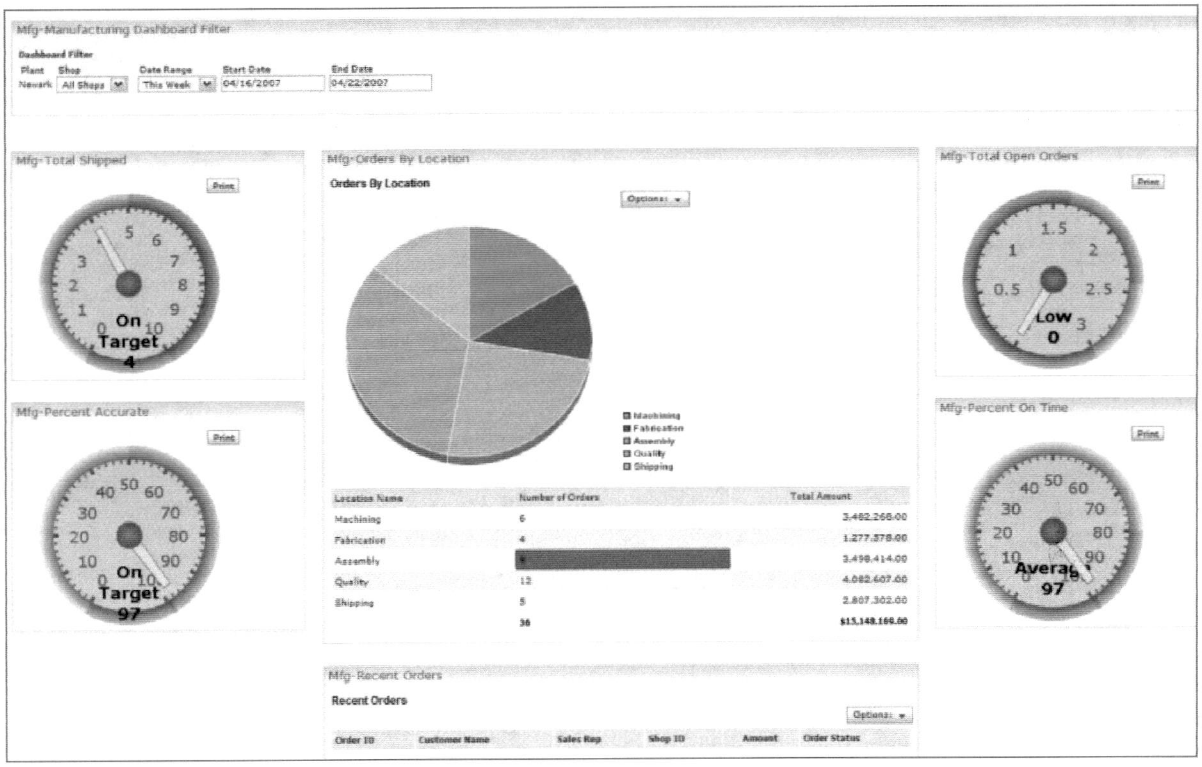

Figure 7 – Sample KPI Graphic 4

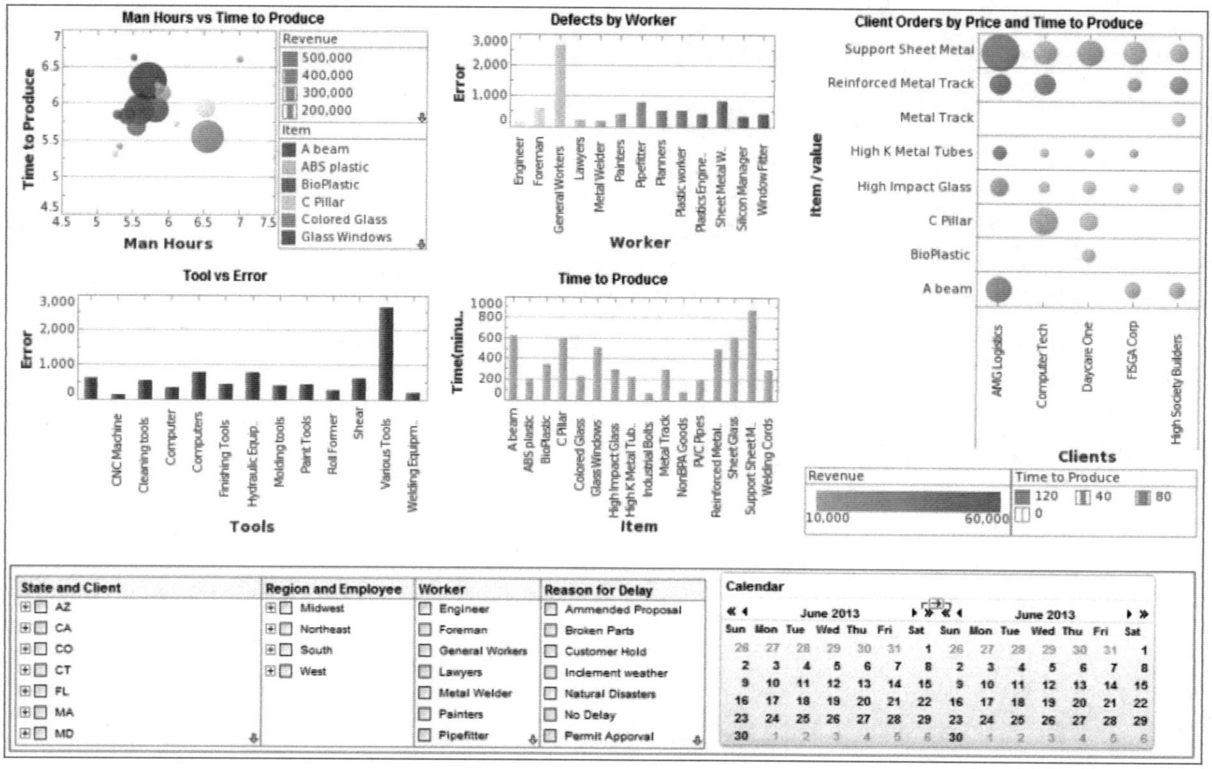

Figure 8 – Sample KPI Graphic 5

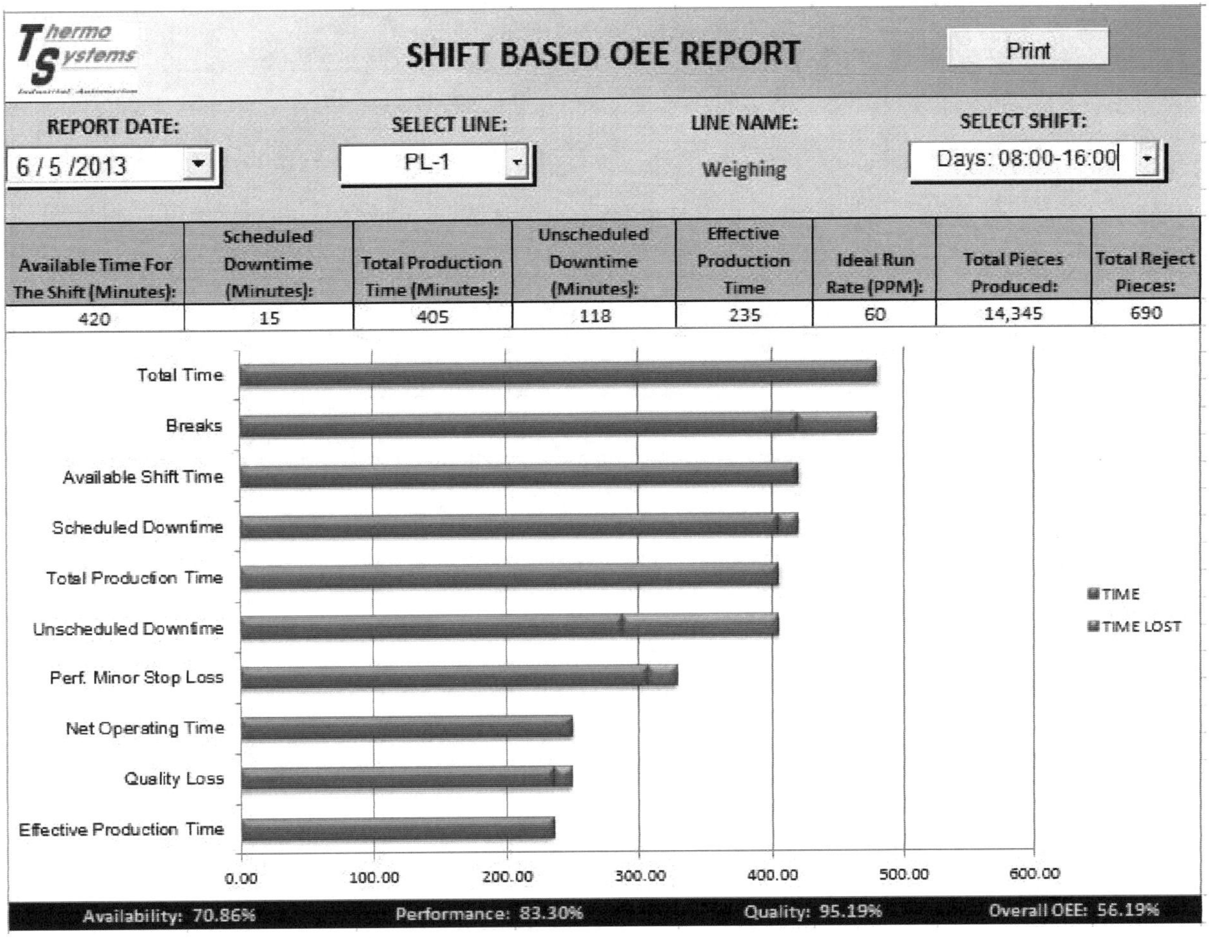

Figure 9 – Sample KPI Graphic 6

Make Up Water Meters	Previous Day Midnight Readings	Current Day Midnight Readings	Daily Consumption/Usage	Time Period To Date Total	Units
Expansion Tanks	363429	363829.4	400.4	#VALUE!	100 gallons
Cooling Towers					Gallons
Main Towers	14081417	14081717	300	#VALUE!	500 cu.ft
Plant Effluent Meters					
Sump #1 (FT/FE 8454)	5278045	5279045	1000	#VALUE!	Gallons
Sump #2 (FT/FE 8456)	4886254	4887254	1000	#VALUE!	Gallons
Energy Sources					
MTCC Gas (main)	814071	815071	1000	#VALUE!	SCF
MTCC Electric (kW)					kW
Turbine Gas (main)	152691	153691	1000	#VALUE!	SCF
Turbine Electric (kW)	6212936	6213936	1000	#VALUE!	kW
Fuel Oil Tank #1	17252	91.7	-17160.3	#VALUE!	Gallons
Fuel Oil Tank #2	18526	97.3	-18428.7	#VALUE!	Gallons
Fuel Oil Tank #3	21365	87.3	-21277.7	#VALUE!	Gallons
Fuel Oil Tank #4	4710	34.4	-4675.6	#VALUE!	Gallons
Retail Area Meter					
Electric meter	9	10	1	#VALUE!	kW
Electric meter	2	10	8	#VALUE!	Demand
Water Softners					
#1	130100	131000	900	#VALUE!	Gallons
#2	23236	27715	4479	#VALUE!	Gallons
#3	29680	34420	4740	#VALUE!	Gallons

Figure 9 – Sample KPI Report

8. Sources

1. Various KPI Example Graphics
   Sai Lo, PE – Director of Engineering – Thermo Systems, LLC
   Various Screen Shots From Deployed Systems
2. Various KPI Example Graphics (reproduced with permission)
   F. Lewis Cain – Managing Director – Control Systems Integrators
   Power Point Presentation Titled: Using Operational Equipment Effectiveness to Meet Business Needs
3. Journal of Applied Systems Analysis (The DIKW Model)
   R.L. Ackoff
4. Four Steps to Turning Big Data into Action.
   Chris Meyer, Tim McGuire, Maher Masri, Abdul Wahab Shaikh.
   http://www.forbes.com/sites/mckinsey/2013/10/22/four-steps-to-turn-big-data-into-action/
5. Systems Thinking: Data, Information, Knowledge, and Wisdom
   Gene Bellinger, Durval Castro, Anthony Mills
   http://www.systems-thinking.org/dikw/dikw.htm
6. The DIKW Pyramid
   WikiPedia,com
   http://en.wikipedia.org/wiki/DIKW_Pyramid
7. Meriam Webster Online Dictionary
   merriam-webster.com
   http://www.merriam-webster.com/dictionary
8. Six Important Stages in the Data Processing Cycle
   Paul Rudo
   http://www.enterprisefeatures.com/6-important-stages-in-the-data-processing-cycle

[END OF WP_AICHe-2015-001]

Abstract ID#402421

CAN THE SCIENTIFIC METHOD SURVIVE THE PLANT CONDITION?

Robert G. Cook, Veolia, Houston, TX

Abstract

If engineering is the application of technology to improve our way of life, then the venerable pilot is the means to that end in the chemical and process engineering realm. Whether a pilot is conducted to develop a new product, assess methods to recover a waste/byproduct, troubleshoot, enhance robustness or improve the quality-yield-capacity of a long established operation, the execution of technically sound pilots is inextricably tied to the Scientific Method (SM). The SM categorically transcends the industry, market or application and establishes a powerfully simple method of gathering data and knowledge to accomplish a goal. That it is applicable to grade school students where it is first introduced may belie its utility to thirty-five year subject matter experts in the various high-tech fields, but make no mistake, those who snub it with technical arrogance do so at great risk to the success of their undertaking.

While the SM's natural habitat is the comfort zone of HVAC-controlled Research and Development (R&D) laboratories around the world, its survival is less certain when immersed in the comparably hostile ecosystem born of the "plant condition". In this paper, the so-called plant condition is the confluence of human-factors, commercial, technical, safety, and financial/resource limitations that conspire to make it a casualty in what may very well have started as a well-planned undertaking with a clear Design of Experiments (DoE). The plant condition can also take on a more literal meaning – when the plant is already in a start-up mode and its performance is lacking in some way, thus necessitating additional work alongside start-up – perhaps even leveraging the plant itself as part of the plant. While piloting alongside a start-up is certainly not a comfortable position for a new plant, it is a scenario that demands the most from the best qualified, disciplined, and creative team members and forces a disciplined adherence to the SM in spite of temptations to deviate from it. A failed pilot is at best a lost opportunity and at worst the harbinger of a failed venture so it may be difficult to locate a post mortem or read of the lessons learned as the victims (not to mention those accountable) often have little incentive to propagate the pain.

This paper takes a very process-centric look at how the SM lies at the foundation of properly executed pilots and how the various forces inherent in the plant condition can be managed to keep their execution on track towards achieving the goals. It demonstrates how the tenets of process safety management, good engineering/project management techniques, and an ever-increasing desire to develop more chemically responsible and sustainable solutions can be used to maintain focus on the priority and goals to help ensure success, while acknowledging that the goals themselves may change as data is gathered and the pilot organically adapts in response. In the final analysis, the raison d'être of the chemical engineer and what generally separates them from other disciplines is the mass and energy balance (MEB) gleaned from a properly executed pilot. Consequently, it is not an overstatement to suggest that the SM is the essential tool that should steer pilot endeavors if a goal is to be met, if a process is to see commercialization and if a business is to succeed. Indeed, if society's way of life is to be advanced.

Keywords
Process Piloting, Chemical Process Engineering, Plant Troubleshooting, Innovation, Scientific Method, Entrepreneurial Business Development, Plant Condition Leadership

Introduction

The modus operandi of innovation, discovery, and chemical engineering/troubleshooting is bound to the scientific method (SM) yet the topic is not afforded any formal treatment in the definitive tome that is Perry's Chemical Engineers' Handbook [1]. Perhaps that is because the SM is an implicit best practice that all diligent Chemical Engineer's (ChE's) already follow. If true, why then does it seemingly fall victim when needed most, such as in entrepreneurial start-ups and the chaotic conditions enveloping problematic processes where the boss is asking "is it fixed yet?" seemingly every ten minutes? While there exists a plethora of on-line and printed articles covering a broad range of pilot-related topics such as sticking to the basics, fast-tracking execution, practical ways to design and reduce time/costs, how to make pilots amenable to scale-up [2-7], far fewer address the comparatively esoteric plant conditions inherent in the fast-paced, chaotic world of entrepreneurial activities that frequently require sub-optimal parallel-path process development and operationalization. The confluence of commercial, operational, process, and human factor forces that conspire to impose stresses on the ChE while caught in such a circumstance make it tempting to dismiss the SM in pursuit of a mutable priority.

Over the past 25 years, the author has had the good fortune of developing a successful entrepreneurial start-up based on an innovative solution centered on creating a valuable resource via leveraging spent caustic solutions as feedstock. That work implemented a combinatorial innovation[1] employing lessons learned from the metals finishing and mining industries coupled with technology common to the chemical industry to yield a solution provided to the refining HF alkylation industry. Suffice it to say, that is a lot of ground to cover for an entrepreneurial startup. Over the ensuing 20+ years, pilot deployments ranging from upstream produced water to complex integrated chemical plants essentially functioning as zero-liquid discharge water processes have been the working environment, or more aptly *battleground,* of yours truly. By virtue of the experience garnered from working in multiple industries of varying size, a fascinating realization is that regardless of how broken a plant might be, how much pressure the client is applying, or how dysfunctional the personalities, conflicting business units, intellectual property disagreements etc. become, it is critical that these "plant conditions" not conspire to dissuade adherence to the SM. If that seems self-evident, then I would restate the assertion with elevated rhetoric: Engineers tend to recognize details immediately but the obvious takes more time.

This paper presents common human factors and plant conditions that can conspire to derail the SM, factors that one typically does not learn about in the classroom or study in the literature. Some aspects discussed herein are addressed in relatively recent books on innovative inquiry [8], lean start-up principles [9] and essentialism [10] and the concepts that match author experience are duly noted. The paper's content and structure are based on a general block diagram of the SM. After all, is it not a best practice to consider the goal, structure and content of a paper before delving into actually writing it? By the same token, all fruitful endeavors of inquiry begin with problem definition.

It is not the point of this paper to dwell on details of the SM, as it is assumed that most readers still maintain a solid grasp of its principles from as far back as the fifth grade. Rather, factors systemic to entrepreneurial process innovations, start-up and troubleshooting are presented along with some recommended approaches to manage them. The paper will aspire to be a technology neutral guide and rely on general examples of recurring phenomena witnessed by the author over his chemical process engineering career.

---

[1] A manner of innovation where technologies, concepts, services, ideas (often familiar), et al. from apparently disparate sources are combined in unique and novel ways to yield innovative solutions. The process is facilitated by divergent thinking outside of the apparent subject matter scope.

The Scientific Method

The author experienced jamais vu[2] on entering a fifth-grade science classroom during a parent/teacher meeting in 2014. Among the scattered exhibits of periodic table artistry hung a comparatively nondescript poster entitled "The Scientific Method" which appeared much like that shown in Figure 1. Stopping for a moment to take in the realization of how early the concept is taught, a cursory review of each block was accompanied with cascading flashbacks of various high-intensity projects that became encumbered in plant conditions which pulled the project away from the SM's elegant simplicity, in some cases significantly delaying resolution of the problem and unfortunately in a few others, resulting in failure of the opportunity. An epiphany emerged on deeper reflection and in each case it could be demonstrated that "We failed to follow the SM!"

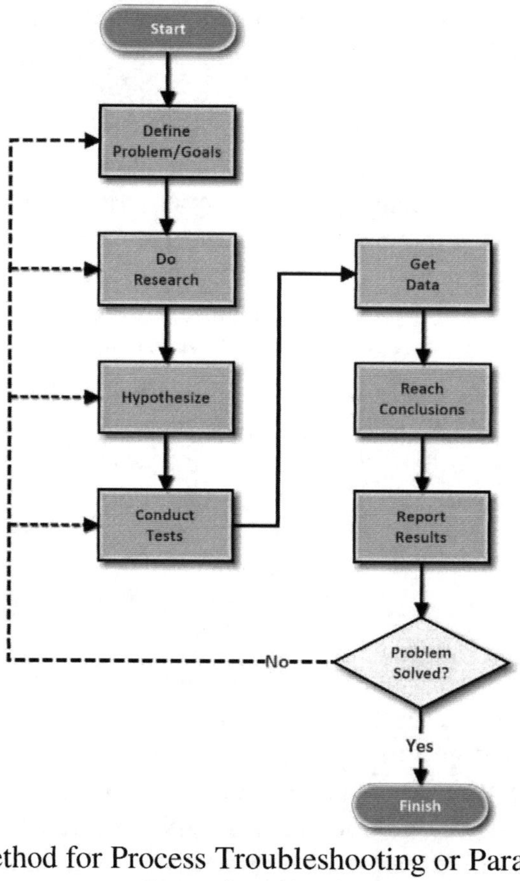

Figure 1: The Scientific Method for Process Troubleshooting or Parallel-Path Development and Operationalization

Steps of the SM

The SM is a proven process used to guide the execution of carefully controlled experiments in a way that is logical, orderly and affords the application of rational/critical thinking skills to solve problems and obtain new information in an unbiased manner. To be termed scientific, a method of inquiry is commonly based on empirical or measurable evidence subject to specific principles of reasoning. While SM diagrams can take on various flavors based on the particular field of study, for

[2] The opposite of déjà vu where something that is very familiar is suddenly seen in a new or significantly different context, making it seem new or experienced for the first time.

purposes here, the key steps of troubleshooting a process and/or acquiring critical information in a parallel-path process development and operationalization are as follows:

1. **Define the Problem** or establish the goals of the work
2. **Do the Research** to learn what is already known
3. **Hypothesize** and develop a path forward based on an educated guess as to what might solve the problem and assemble equipment, materials, plans needed to proceed
4. **Conduct the Pilot Work** to test the hypothesis
5. **Get the Data** from the tests, both operational and analytical
6. **Reach Conclusions** based on a rational analysis of the data
7. **Report the Results** and determine if the problem is solved or goals have been achieved

When the conclusions indicate the problem remains or the goals have not yet been achieved, it is necessary to re-enter the SM flowchart at potentially any point from redefining the problem down through conducting additional tests. Each project and circumstance is different and it is not the goal of this paper to assess the nuances of every particular scenario but rather to point out the importance of following the SM in every case.

Criticality of Problem Definition and Setting Goals

Einstein has been quoted as saying "If I were given one hour to save the planet, I would spend 59 minutes defining the problem and one minute resolving it". Similarly, by far the most important step in the SM is to define the problem/goals first. When there exists a lot of pressure to get the process working properly, there is often a rush to begin testing ideas or jumping to conclusions before any objective data is obtained. This is symptomatic of a malady the author refers to as technical arrogance which is remarkably chronic even as the results suggest it contributes to the problem, not help resolve it. Subject Matter Experts (SME), chief technology officers, recognized process experts, and especially business leaders with the most invested seem to be at greatest risk of contracting this condition which worsens as pressures mount. Extrapolation is risky to begin with but extrapolating from pure conjecture is a losing proposition, pure and simple. It is missing the characteristics most desperately needed at the outset of any serious undertaking, namely patience and humility derived from healthy self-awareness.

Years ago, in the midst of a particularly frustrating study where the team did not devote sufficient time to identify the problem and set forth the goals, a diagram similar to that shown in Figure 2 was circulated as a stress reduction joke to highlight the seriousness of the situation we had become mired in. The comedic value transformed into stark realization as it became apparent how true it was.

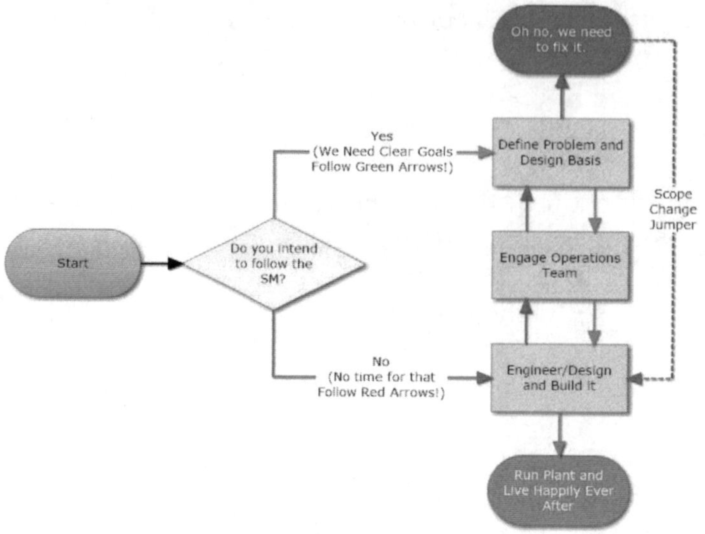

Figure 2: Project Flow When Following (or not) the Scientific Method

As shown in Figure 2, the red arrows are the direction a project takes when the problem and goals are not established at the outset whereas the green arrows are the desired result when they are. Projects can unfortunately get caught up in an infinite loop flowing in the direction of the red arrows. Design-Build engineers and construction contractors make great livings when a project follows the red arrows because each scope change, each new operating procedure revision, and each new "these new setpoints should work" can sap a company dry of capital resources and more seriously, lead to a demoralized team. The situation is similar to the sunk cost fallacy[3] in that both seem to feed on themselves, growing in the perceived need to persist in spite of overwhelming evidence to the contrary. Perhaps psychologists could weigh in on this interesting parallel but the key point is that time must be taken to understand the root problem and establish clear goals before any physical work begins. Some of the plant conditions that might divert attention from first defining the problem and establishing goals are covered below.

Problem Definition and Establishing Goals

Trouble at the Mine

Friday afternoon was coming to a close for the process group in Chicago. Folks were cleaning up, shutting down computers, and chatting about planned weekend activities when Frank's (the process group leader) phone rang. It was Dennis, the operations manager in the western zone. Apparently, over the last several weeks, operations was combating trouble with a growing emulsion layer in the extraction unit and were doing everything they could to deal with the issue. None of the troubleshooting procedures were effective so they had started purging the rag layer to a local pit. The pit quickly filled, so they ordered mobile tanks to store off-spec material hoping the emulsion would eventually break so that it could be recycled back into the unit. Unfortunately, that was not happening and the mobile tanks were nearly full. To make matters worse, someone called the anonymous hotline complaining that vapors in vicinity of the open pit were making them nauseous and the Mine Safety and Health Administration (MSHA) was on its way to investigate.

[3] Also known as the Concorde fallacy based on the failed commercial airline of the same name, the idea that investment of funds should continue on a project, product, etc. to avert waste of funds already invested and lost which only leads to further losses and more bad decisions.

They had done what any good operations team would do –fight the problem until they were out of options. And they were out of options at five minutes to midnight.

"Can you help us, Frank?" Dennis blurted with an urgency in his voice that was palpable. "I don't know, Dennis. What's the problem?" muttered Frank as he tried to come to terms with the seriousness of this situation. "I just told you Frank. A heavy emulsion layer is building in the extraction unit." Dennis reaffirmed. "We need to get the chemistry adjusted or figure out what kind of emulsion breaker we need to deal with this stuff. It's ugly and it didn't do this in the pilot so we are in unchartered territory here", Dennis went on to say. "So get the development lab lined up to run some jar tests because we only have about 24 hours before someone will need to call Brutus (the business unit head) and let him know we have to shut down. We have been fighting this for weeks and are doing the best we can but the process does not work so that call is going to have to come from you Frank.", Dennis said with an aura of resigned indignation. "Get on a plane and get out here ASAP" Dennis ordered. "I gotta jump off this call and head down to the chemical plant now. The acid they are sending us is all over the map and that's just another issue we've been contending with." Dennis threw in almost as an afterthought. "OK Dennis, I'll be there first thing tomorrow morning." Frank yielded.

The wagons were already circling and Frank knew his weekend on Lake Michigan was turning into Days Inn at the desert. He was tempted to call the lab as Dennis instructed but he realized he did not know what to tell them. He did not even know what the real source of the problem was. Maybe it was upstream in the chemical plant? Dennis did mention the acid quality was off but what did he mean by that, exactly? Perhaps operations folks were simply not following procedures again. They did tend to take liberties when they thought they knew better. Or maybe it was something else altogether? Suddenly, a wave of anger overtook Frank as he realized, "...and Dennis waited until the last damn minute to call? Great!"

The foregoing fictitious scenario has all the makings of a chaotic plant condition gearing up to derail a logical SM, including:

- Operations fighting a problem as it grows to the breaking point
- Alleging what is needed to fix the problem before the root problem is identified
- Defensive posturing and blame
- Implicitly suggesting there is no time (to do things right)
- Flaring emotions that work counter to cooperative teamwork

It is unlikely that Dennis would have been receptive to Frank insisting for a meeting to explore the root cause and establish goals to resolve the problem, not at that particular moment anyway. Therefore, on the plane ride over, Frank went to work. Let's explore some of the issues he considered as he prepared for a meeting on arrival at the mine.

Strict Enforcement of Management of Change (MOC)

Despite ostensibly wise advice to the contrary, symptomatic treatment of problem plants is often necessary to keep them operational in the short-term so that a long-term solution can be developed. However, short-term solutions can and will morph into long-term patches that only delay and intensify the problem. Tanks and pits fill faster than expected and the costs associated with third party disposal services can consume budgets quickly.

In our fictitious example, operations bought time with tanks but apparently did not keep anyone informed of this activity. They made fundamental changes to the operating procedures and that probably introduced new, unintended process risks and possibly other hazards. Regardless of whether a site is under MSHA or OSHA jurisdiction and/or subject to Process Safety Management (PSM) [11], in the author's opinion, adherence to Management of Change where operating procedures are concerned is always a best practice that should be followed.

The moment operations felt the need to purge a rag layer to a pit or bring in temporary tanks, this should have prompted a MOC review. A discussion of MOC is beyond the scope of this paper but this author would suggest that any company wishing to improve safety and operational robustness should consider the tenets of PSM even when they are not strictly required to do so, and MOC is arguably one of the more important tenets of PSM.

While it is easy to lay blame on operations in this example, the process group also failed to stay apprised of the mounting problem at the mine. In small entrepreneurial companies, little happens without all interested parties knowing it. However, in large companies, distance and hierarchy can result in the sentiment "no news is good news". Thus, for projects that have yet to be confidently lined out and operationally optimized, ongoing communications between the operations and process group should happen on a routine basis. This may be in the form of sharing operating logs and status reports, or weekly report summaries. These reports serve as fair warning of problems or, at the very least, are an excellent way to understand recommendations the operations team has to improve the process and to do so in a manner that will not introduce new hazards or risks on the business.

Is the plant in a transient start-up mode?

When in search of a root cause, ask the question "Are any of the upstream unit operation/processes having their own set of problems?" If so, focus on getting the source operation(s) right before concentrating on downstream symptoms. In complex plants, operations can form tribes and these tribes tend to focus on their domain at the expense of others who may be impacted. It can be a survival-of-the-fittest mentality when a lot is on the line in getting a new business off the ground or a novel and innovative process operational. Pressure comes from all directions and this tends to cause tribes to retreat into comfort zones and look at other areas as the source of the problem, not as part of the team to help solve the problem.

Plants are becoming increasingly more integrated

Zero-Liquid-Discharge (ZLD) is not just for water management anymore. The source problem can be buried deep anywhere in the process. In fact, it can be very difficult locating the source problem because the plant appears much like a self-referential (recursive) spreadsheet error. It is tempting to call the most egregious symptom "the problem" when there exists a much simpler and easy to fix cause elsewhere in the unit. The author has seen a small blowdown valve accidentally left open thwart resolution of a mass balance disparity in a large cooling water system for weeks. A quarter turn on a ¾" ball valve and the problem was solved.

IP Struggles may Confound Root Problem Identification

The source problem may be a confidential client operation or involve Intellectual Property (IP) that key team members are not made privy to. Worse, IP from external parties offering solutions may risk the company's position in the market, especially where proprietary know-how (rather than patents) comprise the novelty of the process and it becomes dependent on the third party solution. It is therefore critical to nix the IP and Non-Disclosure Agreement (NDA) provisions early and open up honest and frank communication channels between all parties as soon as possible. Generally speaking, the chances for resolving IP issues is inversely proportional to the number of parties involved [13]. Assuming the IP issues can be resolved, solutions often present themselves when disparate groups become engaged and openly share notes regarding each other's technology and dependent variables. Simple changes to an operation may then reap great rewards on resolving serious problems quickly.

Innovative Inquiry into Problem Definition

Assess the problem from all angles before fully defining it. Do it by asking simple and seemingly fundamental questions. The 'five whys' method of inquiry is a powerful means of getting to the root problem [8]. While it is not the intent of this paper to discuss ways to establish root cause through innovative inquiry, here are a few good points to get started:

> *"A problem well-stated is half solved" - Charles Kettering, Inventor, Engineer and Businessman*

- The five whys of innovation – a powerful tool developed by Saikichi Toyoda (a father of the industrial revolution) and now a common Six Sigma and ISO 9001 tool
- A seemingly simplistic mechanism to thwart the psychology of personalizing cause and effect (human factors) and arrive at the true (root) systemic issue
- Thwarts the tendency to think lazy, i.e., arrive at the apparently obvious cause initially (typically wrong)
- Not limited to five whys and may only take three in some cases (# is not strictly important).

The goal of using five whys is to mine the root cause, to define the problem and clarify the goal. Therefore, when engaged in five why analysis, it is important to:

- Get frank feedback. Is the problem technology, process or operational in nature? This is not a time to find fault with individuals but rather one to establish the facts.
- Are business drivers affecting the decisions - pressuring operations or technical to take risks or run into unknown zones? Operations can be eager to please and this can lead to unintended issues long-term. Business Development (BD) folks may not realize their requests take the plant out of the comfort zone.
- Are we thinking "outside the box"? – A common metaphor referencing the desire to avoid a cognitive bias known as functional fixedness [12]. Functional fixedness limits the use of tools, objects, processes in the traditional manner rather than apply them in new or creative ways. Functional fixedness results from holistic processing, i.e., viewing systems as a whole versus a sum of their parts.

A simple five why analysis applied in a process and operational context is:

1. ***Why did the client call us?*** *Because they have a stream that is causing plugging problems in their filter system.*
2. ***Why are the filters plugging?*** *Because there is some calcium/barium hardness precipitating.*
3. ***Why is the hardness precipitating?*** *Because the anti-scalant is breaking down and it is allowing the metals to precipitate out of solution.*
4. ***Why is anti-scalant being used?*** *Because the client wants to protect their own operation from the same thing.*
5. ***Why don't we start there then?*** *Good idea, let's call the client for a visit to the source operation to learn more about their scaling problems.*

Tempering Entrepreneurial Exuberance and Problem Avoidance

While the preceding has focused on existing processes that require troubleshooting or optimization, a similar approach toward problem identification and goal setting can be used for new processes that are moving from R&D to pilot phase. For example, some hard questions that need to be answered to temper entrepreneurial exuberance, include:

- Does the reaction warrant investment in a process and does this process warrant investment in a new business? Technical zealotry can cloud good business conduct.
- Do we understand the market and is there a clear need or is this just an exciting reaction/observation that can be used to improve an existing operation? An example of broken research is "just get us some samples and we will evaluate the market based on them", or "competitor A is doing something and we have to do it too" even when it isn't clear if competitor A is actually doing anything that makes sense, let alone making money.
- What are the projected Capital and Operational Expenses (CapEx/OpEx) and how does a five or ten year Distributed Cash Flow (DCF) look under the assumed worst case scenario?
- Where are the sensitivities and risks to the process, and are they sufficiently offset by the upside of success?
- Where is the capital coming from and how can we protect the idea from the ravages of an eager market should we succeed? Raising capital typically involves a lot of so-called "forward looking projections" and when the reality of compounded, idealistic estimates lands, it is better that this realization happen prior to construction. The author has seen many internal estimates by creative technical folks not withstand the scrutiny of independent review by a factor of two or more.

Establish Clear, Quantitative Goals

When it is felt that the root problem has been identified, move to establish clear, quantitative goals. A poor mission is one that has nebulous and/or relative goals that can never actually be achieved. Interestingly, a lot of company mission statements are peppered with banal, feel-good language that can never actually be achieved because they lack a measurable goal. Some awful goals the author has been given include:

- Create a higher value proposition in this application
- Increase our innovative solutions offerings
- Engineer us some more revenue/margin on this project

Do the Research

Once it is felt that the root problem is identified and goals are established, the next key step in the SM is to conduct research. The Internet and its competing search engines are turning knowledge into a commodity yet implementing that in creative and innovative ways still defies rigorous methodologies or best practices. When a novel process requires troubleshooting or refinement, it's important to draw on all sources before deciding that additional tests and development are even necessary. While it may seem odd that culling all pertinent information should be difficult, here again pesky plant conditions, e.g., the need to guard business unit IP and the "power in information" mentality can make mining what is already known very difficult.

Back in the technically prehistoric 1990's, dawn of the Internet era and subsequent spawn of its so-called things, two page executive summaries graced the front of lengthy technical reports and were intended for busy Execs who did not have time for pesky details. Nowadays, attention is lost in email messages that are more than a couple paragraphs long. Indeed, a substantial amount of corporate communications are now performed by short text message quips read in between flights or at red light stops. Aside from the conventional routes of proper literature searching, novel processes are largely empirically driven know-how buried deep in spreadsheet details, process models or the aged SME's and veteran operators. Bridging the eclectic mash-up of scientists, engineers, operations and SME's that are typically involved in an innovative process, here are some tips on doing some unconventional research.

An extensive pilot was conducted and a lengthy report was written, right?

It may have taken months and potentially hundreds of thousands if not millions of dollars invested in that final report. Here is a question; did anyone actually read it before the plant was built? Not just skim the executive summary and highlights and then move on to design the plant, but to actually dissect the work and understand the nuances of what was done? Too often reports are monuments to an effort that does not get deserved attention after release. Process engineers are prone to repeat design mistakes when they are disconnected from the R&D groups upstream of the design, not to mention the downstream operations left to contend with and fix their errors. Go back and review the report(s) along with all the source information upon which they were based including the lab notebooks and pilot operational logs.

Sometimes important details never make it into reports

Sometimes, the most important aspects of a development and/or pilot are not what was written in the report but what was left out of the report. Complex teams of disparate and divergent backgrounds can foster political discord that suppresses negative findings, ignores folks based on their academic background or perceived competence, etc. Therefore, make a point to talk to the operators of the plant and those that operated the original pilot, the scientists that did the initial R&D and hopefully helped validate the concept in the lab, including the engineers who were heavily involved but did not (for whatever reason) contribute substantively to the final report/final design. Some examples encountered defending reasons why key results and/or conclusions were left out of a report include:

- We didn't include that test because [insert reason it failed for reasons that were explained away rather than accepted as a valid result]
- They failed to mention *that* run worked but the six runs before it did not. Do you want to know why?
- We did not feel the need to pilot that aspect of the plant because we already knew that part works fine.

The missing link in the broken process chain may be obscured by folks afraid to come forward with new, critical information that was previously unreported.

Is the Process Safety Information (PSI) Current?

In PSM, PSI includes block diagrams, Process Flow Drawings (PFDs) (with a clear mass/energy balance), Piping and Instrumentation Diagrams (P&IDs), and operating procedures needed to operate the plant safely and effectively. It also helps to have the detailed process and functional control descriptions and at least a listing of major equipment specifications. This is the plant definition and it's not *just for safety*! In the quest to cull all of these, be wary of the following feedback that might help direct attention to the root problem:

- Folks upstream don't share with us
- Engineers are always changing the recipe or procedure on us so those procedures are worthless

- We started doing it this way weeks/months ago because it seems to be working. No, an MOC was not completed
- Those PhD's don't understand the reality of the plant operation so we just do it our own way
- There is no clear understanding of the holistic plant operation; that mass balance isn't close to being right

Further inquiry into the state of the process information may yield the following additional evidence of broken management and dysfunction:

- I was told to stop asking questions and just do it
- I'm not really sure why we do it this way
- We will get to that [pick critical process/safety need] when we have more time, enough money or it gets approved

Scout for the chronic condition of technical arrogance (a malady discussed above) where assumptions are concerned. "We didn't pilot that part because we already know how that works from the last project." Technical extrapolation from disparate bases is perhaps the biggest contributor to process risk where innovative new plants are concerned. Some of the lowest hanging fruit in unconventional research is identifying the flawed assumptions made by the technical experts and SME's.

Was the plant designed properly?

This question is not referring to best practices, robust standards, redundancy, and judicious value engineering but rather - does an audit of the various major equipment specifications, design bases, control logic requirements match up to what was actually delivered and installed? Common examples include gross design/fabrication errors, egregious value engineering liberties, construction management changes intended to cut costs, ergonomically unworkable layouts, etc. Closely related to these, and reflecting again on PSM tenets, was a Pre Startup Safety Review (PSSR) done and how were the identified items from the Process Hazards Analysis (PHA) addressed? While PHA's are not intended to succumb into process design sessions, a typical Hazards and Operability (HazOp, with emphasis on the Op portion) analysis will reveal weaknesses that may be debilitating to an otherwise excellent process if not rectified prior to start-up.

Develop a Hypothesis from a Structured Team

Since this paper is a technology neutral focus on conducting either a new pilot in an entrepreneurial business or a troubleshooting pilot alongside a full-scale process, instead of defining a sample hypothesis based on a fictitious root problem and research, let us instead look at the organizational structure needed to develop the best possible hypothesis from all team members. In general, the goal is to establish a robust structure that can be used to maintain order throughout the duration of the pilot while gleaning the best input from all members.

It is reasonable to assume that most entrepreneurial undertakings are comprised of a diverse and gifted team of individuals that have all contributed to the project - taking it from concept to reality. That it may encounter problems or still needs to be refined and integrated into a holistic operation that was unable to be tested at the outset remains the final gate to success. This is a time when plant conditions can reveal themselves in full force. From technical egos to panicky commercial and business leaders, the gifted team can mutate into a cross-dysfunctional team if not properly managed. Generally speaking, the more senior the business leader and the more blood/sweat/tears he/she has invested, the more likely their personal ties will cloud objectivity and thwart a deliberate and methodical effort based on the SM. Hence, managing the input to arrive at usable and successful output is critical.

Based on the author's experience, small teams of gifted individuals with strong adherence to scope generally function best in a matrix-type organization where there is little need for a hierarchical structure (see Figure 3). As Thomas Edison famously quipped, "Hell, there are no rules here - we're trying to accomplish something" the concept of a bureaucracy-free development is an enticing, even romantic aspect of an entrepreneurial undertaking and is often the catalyst for those who wish to chase and live the dream of having their own successful business. In large organizations/projects where the members are necessarily much larger and spread out (geographically and from a business unit perspective), managing the team in a more structured manner is crucial. Figure 4 illustrates one approach the author has found fruitful towards grouping and managing an otherwise chaotic flat-line structure. The three pillars of the structure are based on Operations, Engineering and Commercial. The development leader is that person chosen from the process group to direct the development effort and arbitrate the path forward. However, the considerations of Operations, Technical and Commercial are balanced on an equal plane to avoid abuse of power.

Regardless of what structure works best for a particular undertaking, the important point is to establish the pilot team and structure before developing the initial hypotheses and plan. The nature of competing BU's and multiple same-level managers can thwart progress and decision making. It's tough to realize this on the fifth conference call two months into the project. Thus, have each BU select a representative/facilitator to channel the collective opinion/input through. Designate a pilot head that will fairly weigh the input and arbitrate disagreement, strive for resolution and help drive the team towards productive decisions that move the project forward toward success.

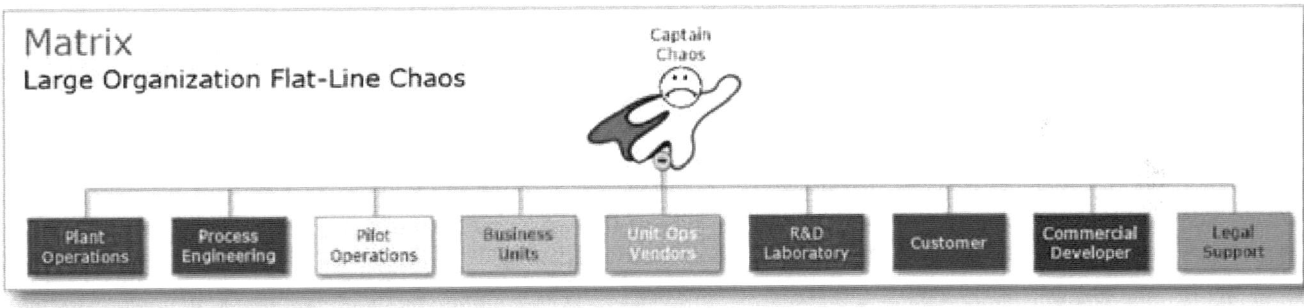

Figure 3: Matrix-type organizations work well with small groups but can confuse resource management in relatively large pilot undertakings

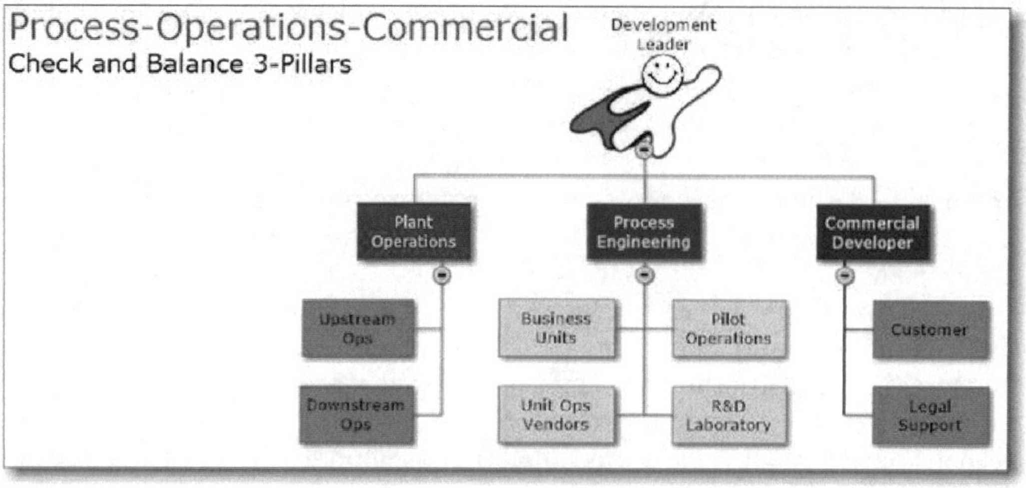

Figure 4: A three-tier structure grouped by Operations, Process and Commercial (BD) generally affords a more efficient means of managing a team assembled to provide process troubleshooting and optimization

Conduct the Tests

The essentials of pilot design, construction and execution have been covered extensively in several *Chemical Engineering Progress* articles [2-7]. Further to the excellent points made therein, the author's own experience suggests that regardless of how bureaucratic or ancillary it may seem, applying the tenets of good process engineering will go a long way towards maximizing the likelihood of a successful pilot outcome. The more complex the nature of the work, and the more diverse and complex the team involved in contributing to the outcome, the more essential it is to focus on proven process engineering practices. Too often, they are ignored in the interest of time or "pilots don't need that". However, it is sometimes more important to have a solid process engineering foundation in novel pilots than it is in the full-scale plant itself lest it never see the light of day. Let's explore the list of tasks all pilots must have for proper and safe execution:

- ***Develop an experimental plan*** - Where necessary, establish a Design of Experiments (DoE) using an outside consultant to facilitate development of a run matrix. Establishing variables critical to the work but outside plant control are often the biggest challenges towards maximizing the value of runs and minimizing wasted resources. For example, if an upstream operation provides unreliable and variable feedstock and is outside your ability to control, it will be important to factor that into the DoE. This is one of the biggest challenges of piloting complex holistic operations.
- ***Develop a process description*** that represents the pilot design. Include block and illustrative flow diagrams that operations staff can reference along with all major unit operations, their purpose and design bases. Photos are especially useful to highlight drawings vs. reality (see Figure 5) and help strengthen understanding of standard blocks typically used in process engineering drawings.

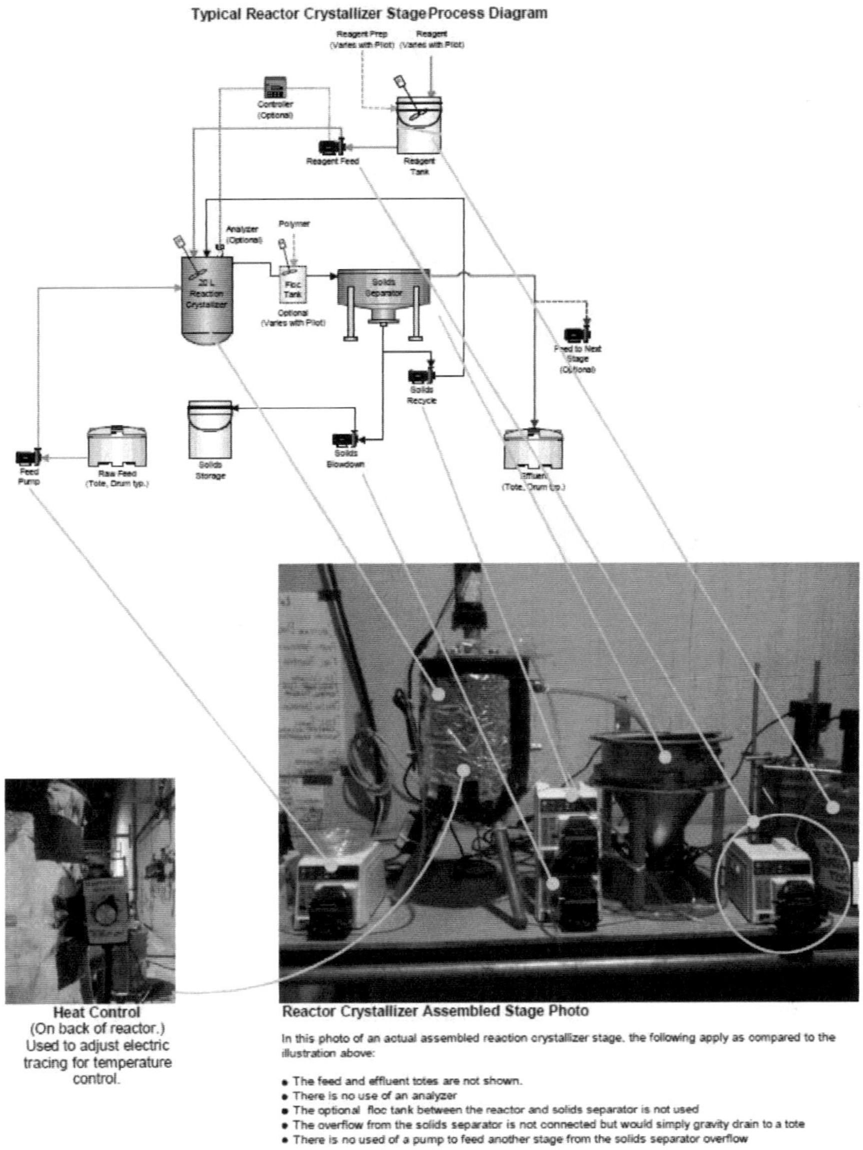

Figure 5: Process Illustration vs. Actual Unit

- ***Label everything in the actual pilot as well as the drawings***. For every piece of equipment, instrument or controller, it is helpful to have a label for the particular setting desired for that run, setpoint and allowed/expected deviation (above and below) the setpoints. Have a 24/7 call-in number if any problems arise that operators can not address by the procedures and training given. Order a case of laboratory label tape and a box of indelible markers.
- ***Generate routine data operating logs*** and pinpoint in photos where and at what frequency data is to be collected. Note the data collection round guidelines and particularly critical variables/equipment to watch. Figure 6 is a very simple example of an operator log sheet. It is intended to keep them focused on rounds and to help identify deviations that might otherwise go unnoticed. Of course, each pilot will have its own logging requirements. Even when fitted with sophisticated Supervisory Control & Data Acquisition (SCADA), operating logs are always good to ensure close monitoring of the operation.

Pilot:
Rx Stage:

Date	Time	Feed (ml/min)	Reagent (ml/min)	Rx pH	Rx Temp.	Recycle (ml/min)	Blowdown (ml/min)	Notes

Figure 6: Simple Pilot Operating Log Example

- ***Develop a Sample Collection Map and Analytical plan***. For plants with their own analytical labs, assume that they will be working hard to provide the plant with what it needs first. The comparatively esoteric needs of a pilot will come second so it is important to have a sample labeling and analytical requirement plan that is intuitive and simple. Of particular importance, note interferences or other components known to skew or create problems with the analytical equipment/procedures typically employed. Figure 7 is an example diagram showing sampling points in yellow boxes on a typical crystallizer pilot the author has been involved in. Figure 8 is a companion Analytical Requirements Plan.

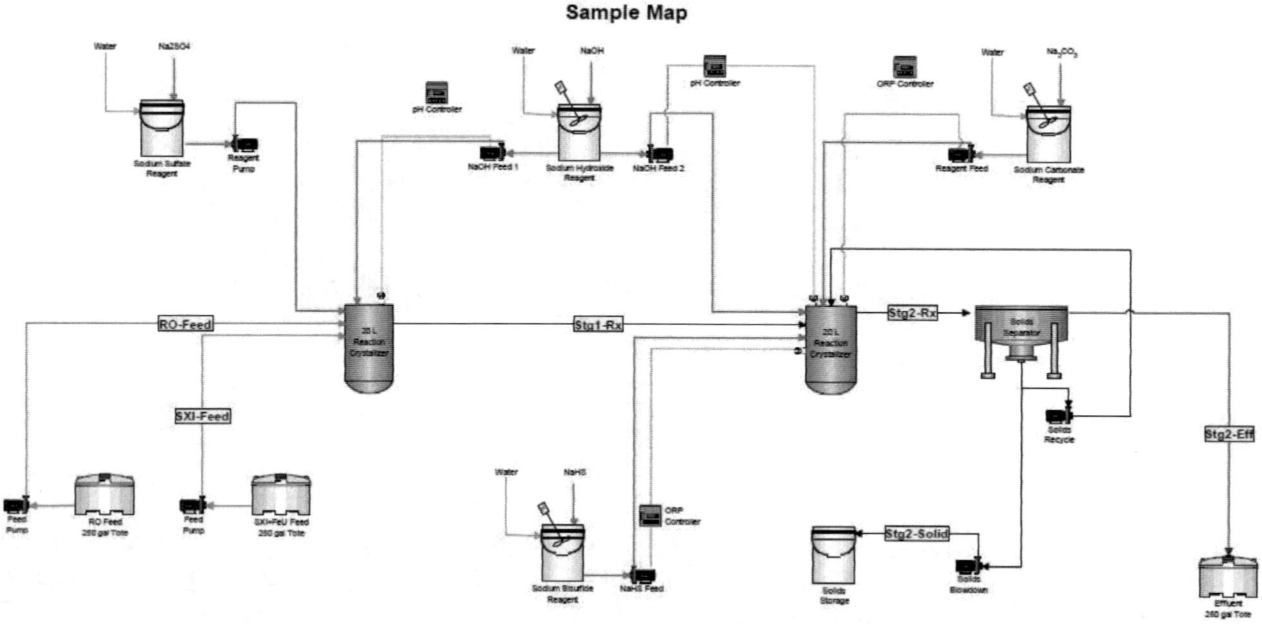

Figure 7: Example Sample Collection Diagram (Sample labels in yellow highlight)

Pilot Analysis Table

Sample ID	ICP-MS	IC	tREO	PSD	TSS	TOC	Alkalinity	Acidity
RO-Feed	X	X					X	
SXI-Feed	X	X	X			X		X
Stg1-Rx	X	X	X		X	X		
Stg2-Rx	X	X	X	X	X	X		
Stg2-Solid					X			
Stg 2-Eff	X	X	X			X	X	

Notes: Conduct Alkalinity on filtered samples to pH 8.3 and 4.5
Conduct Acidity on filtered samples to pH 4.5, 8.3 and 10.7
IC and ICP analyses on filtered samples
Note that tREO in SXI Feed can be up to 5 g/L. At Stg1-Rx (and downstream), tREO is typically < 500 ppm
PSA on raw samples (unfiltered)
ICP-MS includes: Na, Mg, Al, Ca, Cr, Mn, Fe, Co, Ni, Cu, Zn, Sr, Ba, Pb
TOC = total organic carbon
IC includes: F-, Cl-, (NO3)-, (PO4)3-,(SO4)2-
TSS = total suspended solids

Figure 8: Example Pilot Analytical Table

When samples are submitted using the above recommendations, the lab knows where it is from and can quickly determine analytical needs and draw upon their experience with other samples from the same location.

- ***Provide pilot operating procedures and training*** for (at a minimum) start-up, shut-down, and routine operation (Follow PSM requirements). Make sure trained operators are certified to operate the unit independently.
- ***Label all reagents/chemicals clearly*** and provide the new batch make-down instructions directly on the containers/totes/buckets.
- ***Provide general SOP's for common activities*** such as chemical reagent preparation. Working with acids/bases is a good example. Ensure that all operators of the pilot are trained and sign off based on classroom and field training.
- ***Adhere to MOC***. Where substantial pilot changes mandate, conduct mini-PHA's to ensure risks are properly identified and ranked. These need to not be formal HazOp/What-If sessions, but rather pre-run meetings in the field to ensure the score-card reflects the changes and new hazards/risks are considered.
- ***Adhere to Job Safety Analyses*** for all pilot work and activities outside of normal, routine procedures.
- ***Conduct a Pre-Startup Safety Review*** for major pilot changes that require PHA's and be sure to address any open items before operation.

In addition to the above, the following items are recommended but not essential:

- ***Include data sheets/references*** to all major equipment employed in the unit, including where to acquire spares, in-kind replacements.
- ***Provide general troubleshooting tables for common problems and corrective actions***.
- ***Camera and/or Video Camcorder*** – subject to security restrictions, pictures and video taken during pilot campaigns are excellent ways to share observations with other team members. Linked to a secure shared cloud service, it makes for almost immediate sharing of real-time status and findings.

Lastly, establish a culture of trust and empowerment by giving all pilot members the ability to stop work. Encourage copious note taking outside of routine data gathering rounds. Encourage operators to spot deviations from past operation campaigns.

Get Data & Reach Conclusions

The two types of data collected during most pilots are operational and analytical. As the saying goes in computer processing circles, "Garbage in, garbage out", the conclusions reached from pilot tests are strongly a function of the quality of the data obtained. Conclusions based on faulty analytical will lead to improperly planned additional tests. Bad analytical can also yield false hopes that the test was successful or conversely, indicate the test failed when in fact it succeeded. Waiting on analytical data can be almost as debilitating as bad data. By the time the data comes back, the window of opportunity to assess other conditions may have passed or the sample volumes/other restrictions may force breaks in operation. Therefore, have an agreement in place with a lab that is able to respond to variable sample campaigns and frequencies.

On the operational side, attempting to operate complex pilots without the right talent can result in failed runs. In times of operational strife, the main plant operators will already have their plates full trying to keep the operation going so there will be little appetite to conduct quasi full-scale pilot tests. If slip-stream stand-alone pilot systems are implemented, pulling these knowledgeable operations folks to help with the pilot would be ideal but more often, the project will have to make do with temporary help, perhaps from other units or worse, from outside agencies where training might begin with a review of the metric system.

A properly managed cross-functional team will help maximize confidence in conclusions drawn so that the path forward can be optimized in terms of value and resource utilization. Here, the plant condition can once again create an issue when, for example, a single set of data agree with a particular high-ranking technical member while a myriad more statistically trounces his/her hypothesis and yet they stubbornly cling or attempt to find fault with how the data were obtained. Analytical chemists or operators can be easy targets who will be reluctant to stand up and defend their work, especially when the technical challenge is the Chief Technology Officer (CTO). Outspoken and aggressive personalities can attempt to suppress the more logical yet undesirable findings presented by passive personalities that do not want to enter into what could be a contentious debate. The author has seen a case where a particular process engineer was asked to be removed from the team because a higher-ranking member did not like his/her reporting the failed runs commingled through a reported successful campaign.

Reaching valid conclusions is the ultimate reflection of a coherent and properly managed team. In view of some of the issues discussed above, below are a few tips that I learned along the way to help reach them faster:

- As noted above, establish an analytical support group that is willing and able to respond immediately to unpredictable sample frequency. One that may need to analyze constituents outside of the plant norm and/or that has interfering compounds that can thwart reliable results. At times like these, a three-day turn-around is not rushed, it is agonizingly slow. Results from IC/ICP should be available in hours whereas more complex results from instruments like a GC should be within 24 hours.
- Staff the pilot with operators pulled from the plant where possible. If not, have routine checks by plant operations to support less experienced operators. Ideally, the lead process engineers and/or scientists involved in the process development and troubleshooting should share quality pilot time 24/7. Never leave third party temporary help alone to run the pilot.

- Schedule conference calls based on clear agendas with desired goals. Manage members to prevent aggressive personalities from overtaking the call. Reach overall agreement on the best path forward even when not all team members agree. Managing by committee consensus never results in swift progress.

Report Results

Pilot results are typically reported via email or on routine conference calls so that the team can weigh in on conclusions and suggested next steps in a live session. One of the most powerful means of sharing results is through a "living mass balance", i.e., one that is updated to reflect results as work progresses. This mass balance need not be one that maintains perfect component/ionic species balance, but rather a high-level empirical summary building on a process block flow diagram that everyone can glean from a glance what the results are. Leveraging tabs in a spreadsheet is a great way the author has found to separate analytical, operational and process results.

When the problem remains unresolved, conference calls are helpful in soliciting creative input to assess where the dotted line in Figure 1 ultimately needs to land. Not surprisingly, conference calls are fertile ground for the plant condition to reassert itself. The two most common reasons include results that do not support the hopes, hypotheses or agendas of other team members or flamboyant personalities that can commandeer calls and take it far off course as they pontificate on nuances that do little that advance the call. Even when conference calls are regularly scheduled, it is important to thwart the plant condition by:

- Create and issue a call agenda prior to each call
- Hold strictly to the agenda and time allotted
- Establish the expected goals of the call at the outset
- Issue a call summary along with the agreed path forward
- Highlight action items and responsible parties so they are accomplished in a timely manner

One of the more difficult aspects of leading a conference call is forcing folks to listen when someone is speaking. Instead of listening closely to the point a speaker is making, many folks will use the time to prepare a rebuttal instead of commenting constructively. In preparing this paper, the author spoke with managers that have become flabbergasted by young engineers that seem to have forgotten the importance of listening. A high ranking manager framed it in this manner, "I never learned anything new listening to myself speak." In leaving one meeting where an individual commandeered the time, a disgruntled attendee flatly stated, "He who says the most seems to have the least to say." It is striking how much can be learned through focused listening.

While internal calls should be open and frank sessions, always conduct the call with respect and proper business conduct. Assume that it is possible unexpected third party attendees could be present on large calls. The author was on a conference call many years ago where the client announced his presence well into the call, after comments pertaining to his management skills (or lack thereof) were called into question by a member of the pilot operation team. Given his personality, it did not create a huge problem but it could have been a serious faux pas for that operator. The moral of the story is that when plant problems persist, personalities will flare and it is essential to maintain respectful and ethical conduct at all times.

Closing Thoughts

Perhaps there is no greater milestone for an aspiring chemical engineer than to participate in a pilot to validate a novel process crucial to an entrepreneurial startup or one intended to troubleshoot/optimize an existing operation at a large corporation that has the eye of senior management. The internet has made knowledge a commodity to the extent that it seems almost trivial to design many pilots in a manner that is cost effective, amenable to change and will actually yield data that can be relied on for a successful scale-up. And where the right tools cannot be found in the literature or online, SME's and process consultants abound to help get the hardware right to meet the challenge. However, while the chemistry and physics are challenges in themselves, harnessing the mash-up of talent can present even greater difficulties. Referred to herein as the plant condition, these include various circumstances and human factors that must be managed to ensure success. As any seasoned manager will confide, managing a process can be easy compared to managing personalities, especially where there is a lot on the line.

This author contends that a very simple framework exists to prevent plant conditions from taking pilot plans off track and that is quite simply the SM. The SM is an elegant and proven process that can be relied on to help resolve external pressure/temptation to deviate from a logical progression to obtain data upon which valid conclusions can be reached, leading to a successful outcome. It may seem academic or even trite to assert that something as simple as the SM, typically learned in grade school, can have such a profound impact on 25+ year veterans and SME's. This is why staying self-aware and maintaining humility and respect are so crucial in times of high-stress development efforts. The SM is not a guarantee that an idea can be brought to market or a problem project can be turned around but this author alleges, through the examples and recommendations presented herein, that it is in fact the best chance to accomplish those worthy goals. In the end, nothing breeds passion, confidence and the desire to tackle greater challenges than success itself. Most chemical engineers of note this author has met are driven by results and nothing compares to being immersed in a results-driven team when there is a lot on the line.

References

1. **Perry, R.H. and D.W. Green, Eds.,** "Perry's Chemical Engineers' Handbook," 8th ed., McGraw-Hill, New York (2008).
2. **Ginkel, S.L. and D.L. Olander,** "Stick to the Basics of Pilot Plant Design," Chem. Eng. Progress, January 1991.
3. **Martin, P. and M.J. Dietrich,** "Fast-Track Your Pilot Plant Project," Chem. Eng. Progress, January 2005.
4. **Dukhedin-Lalla, L.,** "Don't Apply Commercial-Plant Specifications to Pilot Plants," Chem. Eng. Progress, February 2005.
5. **Leng, R.B.,** "Scale Up Specialty Chemical Processes Directly," Chem. Eng. Progress, November 2004.
6. **Berg, D.A. and M.T. Cleary,** "Reduce Piloting Time and Cost," Chem. Eng. Progress, July 2011.
7. **Tanthapanichakoon, W.,** "Accelerating Process and Product Development," Chem. Eng. Progress, February 2013.
8. **Berger, Warren,** "A More Beautiful Question: The Power of Inquiry to Spark Breakthrough Ideas," Bloomsbury USA, New York, NY (2014).
9. **Ries, Eric,** "The Lean Startup: How Today's Entrepreneurs Use Continuous Innovation to Create Radically Successful Businesses," Crown Business, New York, NY (2011).

10. **McKeown, Greg,** Essentialism: The Disciplined Pursuit of Less. Crown Publishing Group, New York, NY (2014).
11. **U.S. Occupational Health and Safety Administration,** "Rule for Process Safety Management of Highly Hazardous Chemicals," 29 CFR 1910.119.
12. **Cherry, Kendra,** "What is Functional Fixedness?" *Cognitive Psychology*, About.com.
13. **Worstell, J.H.,** "Overcome Organizational Snags to Technology Transfer," Chem. Eng. Progress, January 2001.

393782 Introduction to High Performance Organizations

Abstract:

In this session we would like to discuss "High Performance Teams - How to best Create and Maintain them!" The topics shall be of interest to a wide range in the audience e.g. from less experienced managers to experts who will share their experience and show effective practices from other organizations. In addition it will assist young students on their way into industry to get a feeling of what they might expect. It is envisioned to discuss effective organizations, hiring practice, company cultures, how to motivate and retain good people, in short: the whole value proposition of highly successful organizations. In order to get the most out of this session good and bad examples shall be discussed.

Copyright © American Institute of Chemical Engineers. All rights reserved.

398488 Lean Leadership - for Performing Teams

Abstract:

Fostering talent, growing the employees, innovating the way teams work – are aspects of a modern organisation that are enabled by Lean Leadership. We will discuss what defines Lean Leadership, why it is successful and how to avoid pitfalls when implementing it. Lean Leadership does not mean less leaders but true leaders as compared to managers. Applying Lean principles to leadership results in performing teams and committed employees

Copyright © American Institute of Chemical Engineers. All rights reserved.

398893 Developing and Retaining High Performing Employees

Abstract:

We will interactively discuss several topics which can play a role in developing and retaining high performing employees.

What gets YOU, the leader, up in the morning? Why are YOU excited to come to work?

How do you communicate this to existing team?

How do you sell this to your perspective employee?

Why should anyone feel the same about your organization?

How does what your team deliver make a difference?

What will everyone learn?

-Offer Mentorship to less experienced people

-Demonstrate calculated risk taking

-Do what you say you will do

-Have their backs

Hire good people and let them work

Who is afraid of their team?

How do other departments view your team?

Your work?

Your importance to the success of the company?

How can NIH* harm you?
*Not Invented Here

Does your team feel comfortable arguing with you or contradicting you?

Copyright © American Institute of Chemical Engineers. All rights reserved.

399765 Making a Dysfunctional Team into a High Performance Team

Abstract:

We typically think of a high performance team as consisting of individual top performers. What happens when you are put in charge of a dysfunctional team that is required to operate at a high performance level? How do you develop the individuals and the team? This presentation discusses the combination of leadership styles and key steps used to move a dysfunctional team into an efficient, competitive unit. It will address some of the significant challenges faced and how they were managed.

Copyright © American Institute of Chemical Engineers. All rights reserved.

400811 Transformational Leadership: Creating a High Performance Team from a Diverse Pool

Abstract:

With globalization, the workforce mix in any organization is shifting to a more diverse pool of employees. The diversity is encompassing, not just in distribution of age & gender, but also in bringing variability in culture, work ethics, knowledge base and standards & norms.

The presentation will address the leadership challenges and opportunities created by such diversity. It will highlight the special skillset of leadership style needed to harness this diversity and creating a unified team who could outperform expectations.

Copyright © American Institute of Chemical Engineers. All rights reserved.

401407 Chemical Mega Projects and High Performance Teams

Abstract:

A mega project is a high value, complex, interdependent, multi-organizational undertaking that creates a game changing difference in the landscape of chemical processing. The bonds that are formed, the lessons that are learned and the impact that is felt spans careers, industries and businesses. There is nothing more exciting.

Game changing mega projects are usually based upon some new process development that enables the cost optimization, capacity increase, or yield improvement on some needed chemical product. Feedstock changes, energy efficiency or new production routes are usually the chemical engineering technology behind a mega project.

A high performance team is built on excellence, confidence and passion. All team members must be comfortable with knowing that all members know how to execute as a team and not just as a group of people. Like a great shortstop and second baseman effortlessly turning double plays, each member of a high performance team knows that when they toss the ball their teammate will be there or as the member on the receiving end of the toss, they can make the play.

Team dynamics are of support and continual improvement. Tasks are organized and decisions are made with the data that is available. There is always a positive build of solutions from the ever present "can do" attitude. Needs are anticipated. Plans are built, but the act of planning is the important effort, not the plan itself. Goals are clear and the pathway forward may have multiple choices. The team is continually adjusting upward to become more effective, efficient and enjoyable.

Leadership comes from the team, but there is a clear leader. The agenda is transparent and the good of the effort and success of the team is the focus. The individual success is measured more in how many "assists" versus how many "goals". Mega-projects are made for high performance teams. Many complex interdependencies are required to be executed with the excellence that a high performance team demands.

Applying high performance team principles and practices will be reviewed through example from a recent $1.2 B effort. Multiple recent examples of another $400 MM effort will be woven into the mix. Tactics applied, lessons learned and some great stories of overcoming the challenges while delivering against demanding project technology, schedule and cost will be presented.

Copyright © American Institute of Chemical Engineers. All rights reserved.

401917 Ethical Considerations for Chemical Engineers

Abstract:

A summary of recent additions to Texas Engineering Rules, including the requirement for a one-time fingerprinting and criminal background check for all engineers.

Discuss the criteria for ethical decision-making as well as the specific ethical considerations mentioned in the Engineering Rules - to protect the public's health, welfare and property.

Address other aspects of employer/ employee principles or values that are important in the engineering profession, such as

- conflict of interest,
- confidential and /or proprietary information,
- gifts and entertainment,
- bribes and kickbacks and
- relationship with clients.

The presentation will include case studies and illustrative examples for discussion and an opportunity for questions.

KEY WORDS: Ethics, licensing, values

Copyright © American Institute of Chemical Engineers. All rights reserved.

401857 Writing a Successful Proposal

Abstract:

For over 40 years, Joe Reynolds lead a project team at Tracor, an Austin-based applied technology firm. Early on, he developed a strategy for leading his team in writing proposals that could beat competition from the top-tier defense contractors. His team's specialty was in designing leading-edge applied technology for the Defense Department.

After a good idea, a successful proposal continues with an evaluation of your company's strengths and weaknesses. Next, you size-up the competition's strong and weak points. As you complete the proposal, you want to describe any fatal flaws that your competition's solutions may have and show how your design avoids them.

KEY WORDS: strategy, proposal writing, winning

Copyright © American Institute of Chemical Engineers. All rights reserved.

401642 Monetizing Your Intellectual Property

Abstract:

Securing strong intellectual property and patent rights is important to protecting your company's inventions and revenue. Licensing this intellectual property can generate additional value. This session will provide practical lessons on procuring and leveraging an IP portfolio, including how to evaluate and improve the strength of your IP portfolio and analyze competitors' IP. Learn how to identify and protect your IP while avoiding common mistakes when monetizing it.

Copyright © American Institute of Chemical Engineers. All rights reserved.

401609 Intellectual Property Trends in Clean Energy

Abstract:

As the clean energy industry has grown in recent years, it has seen a significant increase in patenting and patent disputes. Data from 2006-2011 show that patent filings in the areas of biofuels, solar thermal, solar PV, and wind energy have been growing at a combined annual rate of 24%, far higher than the global average of 6% during this period. This makes it essential to adopt a strategic IP plan to ensure protection of your inventions and avoid competitor IP. During this session, we will explore trends in patent protection in the clean energy industry and consider how to evaluate the strength of your company's patents, minimize risk of infringement, and block competitors. In addition, we will review recent legislative changes under the America Invents Act ("AIA") that provide new routes for challenging third party patents should a potential risk arise as the clean energy patent space grows more crowded. Specifically, we will address how to use the new Inter Partes Review ("IPR") procedure to attack competitor patents.

Copyright © American Institute of Chemical Engineers. All rights reserved.

401628 Strategies for Leveraging the Inter Partes Review ("IPR") Procedure Prior to Product Launch

Abstract:

To ensure a successful introduction of a new product, it is essential to understand the patent landscape, minimize the risk of infringement, and avoid or challenge potentially blocking patents. In this session, we will explore recent legislative changes under the America Invents Act ("AIA") that provide new routes for challenging third party patents should a potential risk arise during the product launch timeline. Specifically, the new Inter Partes Review ("IPR") procedure to attack patents has proved deadly. In the last year almost 95% of patents that entered IPR did not come out the other end. In this session, we will address how to use IPR to your advantage during product launch, as well as how to maximize the chances that your own patent survives an IPR attack.

Copyright © American Institute of Chemical Engineers. All rights reserved.

401918 Bridging the Cultural Gap in International Business – At Home and Abroad

Abstract:

Working with management and employees requires more than simply 'speaking the language". Dr. Kelm will describe several cultural issues that often come up when doing business in Latin America or China. Similar problems also arise in European business settings. Cultural issues may arise in the context of language, environment, social organization, context, authority, non-verbal communication and time perception.Several case studies, based on interviews with Latin American and Spanish professionals, will be raised for discussion and problem solving.Whether you are planning to work for an international company, relocate to another country for work, or build a successful and effective international team based in the US, you will learn practical strategies for bridging the cultural gap.

Copyright © American Institute of Chemical Engineers. All rights reserved.

AUTHOR INDEX